RAPID REFERENCE

SECOND EDITION

Influenza

Jan C Wilschut
PhD

Janet E McElhaney
MD, FRCPC, FACP

Abraham M Palache
PhD

Foreword by
Klaus Stöhr

MOSBY
ELSEVIER

Edinburgh London New York Oxford
Philadelphia St Louis Sydney Toronto 2006

MOSBY
ELSEVIER

First published 2005. Second edition 2006.

Cover image courtesy of Drs Takeshi Noda and Yoshihiro Kawaoka with kind permission.

EAN: 978-0-7234-3433-7
ISBN: 0-7234-3433-6

British Library Cataloguing in Publication Data
A catalogue record for this book is available from the British Library.

Library of Congress Cataloging in Publication Data
A catalog record for this book is available from the Library of Congress.

Notice
Knowledge and best practice in this field are constantly changing. As new research and experience broaden our knowledge, changes in practice, treatment and drug therapy may become necessary or appropriate. Readers are advised to check the most current information provided (i) on procedures featured or (ii) by the manufacturer of each product to be administered, to verify the recommended dose or formula, the method and duration of administration, and contraindications. It is the responsibility of the practitioner, relying on their own experience and knowledge of the patient, to make diagnoses, to determine dosages and the best treatment for each individual patient, and to take all appropriate safety precautions. To the fullest extent of the law, neither the Publisher nor the Editors assumes any liability for any injury and/or damage to persons or property arising out or related to any use of the material contained in this book.
The Publisher

The Publisher's policy is to use **paper manufactured from sustainable forests**

Printed in China.

Biographies

Jan C Wilschut PhD is Professor of Molecular Virology in the Department of Medical Microbiology at the University Medical Centre, Groningen, The Netherlands. His main research interests include the cell entry mechanisms of enveloped viruses, influenza virus and alpha- and flaviviruses in particular, and viral and tumour vaccine development. Based on his interest in viral membrane fusion processes, he developed the virosome system as a basis for novel vaccine modalities. He is Co-ordinator of the Netherlands Influenza Vaccine Research Centre (NIVAREC) and has published over 150 papers in international peer-reviewed journals.

Janet E McElhaney MD, FRCPC, FACP is Associate Professor in the Center for Immunotherapy of Cancer and Infectious Diseases at the University of Connecticut, Farmington and Advisor to the European Scientific Working Group on Influenza. She holds the Allan M McGavin Chair in Research Geriatrics at the University of British Columbia, Vancouver. Her research focus is on the aging immune system and vaccine preventable diseases, including influenza. Her current research includes trials with young and older adults to evaluate the role of vaccines, antivirals and immunomodulators in influenza and related acute respiratory illnesses.

Abraham M Palache PhD is Global Medical and Public Affairs Director for Influenza Vaccines at Solvay Pharmaceuticals, Weesp, The Netherlands. He was Influenza Research Fellow at the National Influenza Centre, Erasmus University, Rotterdam where he received his PhD. He is on the Executive Committee of the European Scientific Working Group on Influenza and is a member of the Clinical and Public Health Working Groups of the Association of European Vaccine Manufacturers. In 2002 he became founding member of the Influenza Vaccine Supply International Task Force.

Foreword

This book is meeting an urgent demand: to provide up-to-date and essential knowledge on influenza to health-care professionals. The authors manage to strike the difficult balance between comprehensiveness and keeping the most important information readily accessible. The book is not only handy in size, with many and excellent illustrations, but it also caters for almost all tastes of health-care professionals, be they general practitioners or hospital specialists.

During the past years, it has been particularly challenging to keep up with the amount of new information in influenza diagnosis, treatment and prevention. The number of annual publications on influenza has almost doubled in the last 5 years. The emergence of influenza viruses in animals with high potential to cause new influenza pandemics in humans opens even more and complex new facets in the influenza universe, and raises numerous questions for health-care professionals. Many of them will be answered in the chapter on influenza pandemics in this book. It is important to read; I am convinced that the real difference in pandemic preparedness globally will be made in hospitals and primary health-care providers. Access to vaccines and antivirals will remain, if nothing fundamentally changes, the privilege of some developed countries; its timely reach to the right people will depend upon the level of preparedness of health systems at all levels, but particularly in the "field". Much remains to be done to reduce morbidity and mortality from influenza and health-care professionals will remain key in increasing vaccination rates. For too long, and not only in the general population, the position has been: influenza – so what?

Nowadays, most practitioners will disagree with this statement, particularly many health-care professionals in hospitals who are involved in diagnosis and treatment of severe respiratory diseases. However, when it comes to using existing preventive measures, opinions appear to differ. Seasonal influenza vaccine remains under-utilized in most developed countries, despite abundant evidence that it is cost-effective and even cost-saving in several age groups. Vaccination rates of less than 50% in high-risk groups in some of the most developed countries in the world are just incomprehensible. Somewhat perplexing is the fact that mandatory vaccination of health-care workers against influenza is still a "taboo" theme in some countries and vaccination rates remain universally low.

Fortunately, knowledge is gradually increasing that influenza is an important respiratory disease with an estimated 0.5 to 1 million deaths annually. Health-care costs due to influenza are well measured and are considerable, mainly due to hospitalizations and treatment of the disease. Health-care workers or other professionals interested in influenza, who want to contribute to reducing its health and economical impact, will find the necessary information for their endeavours in this book. Its impact will be measured against future trends in vaccine uptake and reduction of death and disease from influenza both in developed and developing countries.

Klaus Stöhr
Special Advisor on Influenza Pandemic Vaccine Development
World Health Organization, Geneva

Foreword to the first edition

Last century, three major influenza pandemics have been responsible for at least 50 million deaths worldwide. In addition, annual influenza winter epidemics have an enormous medical and economic impact on society. It is widely documented that serious morbidity and mortality occur among the elderly. Also, influenza frequently leads to complications in young children. Yet despite this, many people, including health-care professionals, underestimate the importance of the disease.

As stated by the World Health Organization, annual influenza vaccination is the cornerstone for prevention and control. Efficacious influenza vaccines have been available for over 50 years. Nevertheless, their potential to control influenza is not fully exploited. In light of the recent avian influenza virus outbreaks and associated concerns about a new influenza pandemic, many public-health authorities stress the need to expand efforts to minimize the annual burden of influenza and to improve pandemic preparedness.

This book provides an overview of influenza, its pathogenesis, epidemiology, the burden of disease and its health economics. It also summarizes available information on the molecular biology of influenza and the emergence of new influenza viruses. It is written primarily for health-care professionals, so they may understand the facts behind increased efforts to implement immunization policies recommended by health authorities in many countries.

The book recognizes the key role of general practitioners in the prevention and treatment of influenza for the benefit of their patients and society. By publishing this book, the authors contribute to international efforts to emphasize the major burden of influenza and the need for better control.

Albert D.M.E. Osterhaus
Professor of Virology
Erasmus MC, Rotterdam, the Netherlands

Acknowledgements

During the preparation of the first edition of Rapid Reference to Influenza, Maurice Gleeson and James Ryan contributed significantly to the text. We do wish to thank them again, as their contributions have remained clearly apparent in this new edition of the book.

Also, with this second edition, we have highly appreciated the cooperation with the Elsevier project team, and we would like to thank in particular project manager Elisabeth Lawrence and editorial manager Alison Taylor for their efforts in bringing this renewed, and again sometimes challenging, project to a proper end on time.

We do thank all those who have given valuable suggestions or comments, or have provided figures or references for the book, including René Benne, Walter Beyer, Robert Bittman, Laura Bungener, André van Eerde, Leo van Elferen, Ted van Essen, Ron Fouchier, Lisya Gerez, Yoshihiro Kawaoka, Jeroen Medema, Kristin Nichol and Takeshi Noda, with special thanks to Anke Huckriede for her comments on Chapter 4 and to Maarten Postma for his contributions to Chapters 6 and 7.

Finally.... we thank our colleagues at work and families at home for their patience.

Jan Wilschut
Janet McElhaney
Bram Palache

Contents

Abbreviations

ADCC	antibody-dependent cell-mediated cytotoxicity
APC	antigen-presenting cell
ca	cold-adapted
CNS	central nervous system
CT	cholera toxin
CTAB	cetyl trimethyl ammonium bromide
CTL	cytotoxic T lymphocyte
DC	dendritic cell
GBS	Guillain–Barré syndrome
GM-CSF	granulocyte–macrophage colony-stimulating factor
Grz B	granzyme B
HA	haemagglutinin
HI	haemagglutination–inhibition
HPAI	highly pathogenic avian influenza
ICER	incremental cost-effectiveness ratio
IFN	interferon
IL	interleukin
ILI	influenza-like illness
ISCOM	immune-stimulating complex
LAIV	live attenuated influenza vaccine
LPAI	low-pathogenicity avian influenza
LT	*E. coli* heat-labile enterotoxin
LTB	B subunit of *E. coli* LT
MHC	major histocompatibility complex
NA	neuraminidase
NAI	neuraminidase inhibitor
NK	natural killer (cells)
NP	nucleoprotein
PCR	polymerase chain reaction
QALY	quality-adjusted life-year
RNP	ribonucleoprotein
RSV	respiratory syncytial virus
RT-PCR	reverse transcriptase-polymerase chain reaction
SRD	single radial immunodiffusion
SRH	single radial haemolysis
TLR	Toll-like receptor
TNF	tumour necrosis factor

1. Introduction

The recent spread of bird flu across Asia and parts of Europe and the Middle East, as well as the occasional infections of humans with a disconcertingly high number of deaths, have caused worldwide concern about a potential new global epidemic of influenza.[1] Are the ongoing outbreaks of avian flu indeed heralding the next pandemic? Although it is difficult to answer this question, there is broad consensus that the emergence of a new worldwide outbreak of influenza is just a matter of time. Literally for ages influenza has been with mankind, and it would be naive to think that as of the 21st century there will be no further influenza pandemics. However, now for the first time, we may have the opportunity to follow the emergence of a potential new influenza pandemic and to prepare ourselves, so that we may ameliorate the burden of the outbreak or perhaps even avert it altogether at the start.

Influenza, or simply "flu", is caused by a virus. It occurs not only in occasional major pandemic outbreaks, but also in epidemics of variable severity almost every winter.[2-4] The term "influenza" has been derived from the Italian influentia in the mid-1300s, indicating that, at the time, the illness was believed to result from astrological influences. Yet, the aetiology of the disease and the explanation for its peculiar behaviour remained elusive. At the turn of the 19th century, influenza was thought to be due to a bacterial infection with *Haemophilus influenzae*. It was not until 1931 that Richard Shope showed that swine influenza could be transmitted with filtered mucus, indicating that the causative agent was a virus.[5] A few years later, Smith and co-workers isolated the influenza virus from humans with respiratory illness.[6]

The burden of influenza for the society, not only from a clinical but also from an economic perspective, is often underestimated.[7] This relates particularly to the recurring annual winter epidemics. Fortunately, since the virus was

Key Messages

- The impact of influenza is often underestimated – annual epidemics are deadly and costly events, and pandemic outbreaks may be devastating.
- Deaths can be prevented and cost savings made by implementation of appropriate vaccination measures.
- Many people who need vaccination do not get it.
- Health-care workers play a pivotal role in reducing morbidity and mortality from influenza by vaccinating people at risk of its complications.
- An increased annual influenza vaccine uptake will not only ameliorate the burden of influenza epidemics but also contribute to an improved pandemic preparedness.

first discovered, efficient means to contain the infection have been developed. Vaccination is the cornerstone of influenza prevention and control (see Chapter 8). Accordingly, the World Health Organization (WHO) has issued guidelines for the implementation of influenza vaccination programmes in individual countries.[8,9] As a result, many thousands of lives are saved worldwide each year. Yet, in many places, implementation of vaccination programmes remains woefully deficient.[10,11] This implies that significant numbers of people at risk of the complications of influenza remain vulnerable to infection and possibly death. Health-care workers are in a key position to promote vaccination coverage and thus to ameliorate the yearly burden of influenza. At the same time, an increased annual vaccine uptake significantly contributes to an improved preparedness for coming pandemic outbreaks of influenza.

The influenza virus

Influenza virions are enveloped particles of spherical or elongated shape, measuring 80–120 nm in diameter and containing a segmented, single-stranded RNA genome (see Chapter 2).[12,13] The influenza virus belongs to the family of the *Orthomyxoviridae*. There are three influenza genera, or virus types, within this family: influenza A, B and C. Influenza

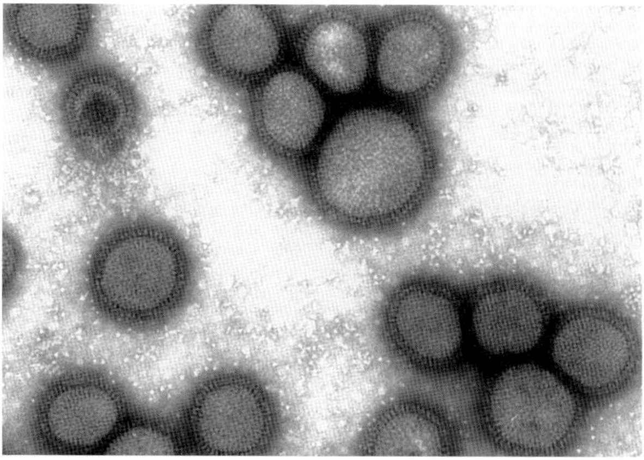

Figure 1. Negative-stain electron micrograph of influenza A virus. Courtesy of Drs Takeshi Noda and Yoshihiro Kawaoka.

A viruses have been responsible for the major pandemics of influenza in the last century and are also the causative agents for most of the annual outbreaks of epidemic influenza. Therefore, this book will be limited mostly to a discussion of the characteristics and impact of influenza A viruses.

Figure 1 presents an electron micrograph of the influenza A virus. A characteristic feature of the virus is its outermost layer of spike-like projections. These are the two viral surface glycoproteins, haemagglutinin (HA) and neuraminidase (NA), which are embedded in the lipid membrane of the viral envelope. HA, the major spike protein, is responsible for attachment of the virus to specific receptors on the host cell surface. HA also mediates a fusion reaction between the viral envelope and the cell membrane, through which the viral genome gains access to the interior of the cell (see Chapter 2). Once inside the cell, the viral RNA is replicated and viral proteins are synthesized, leading to the production of many thousands of new virus particles per cell. Ultimately, the cell dies as a result of the infection. In the lungs and airways, this process of cell lysis leads to desquamation of the respiratory epithelium as one aspect of influenza pathogenesis (see Chapter 5).

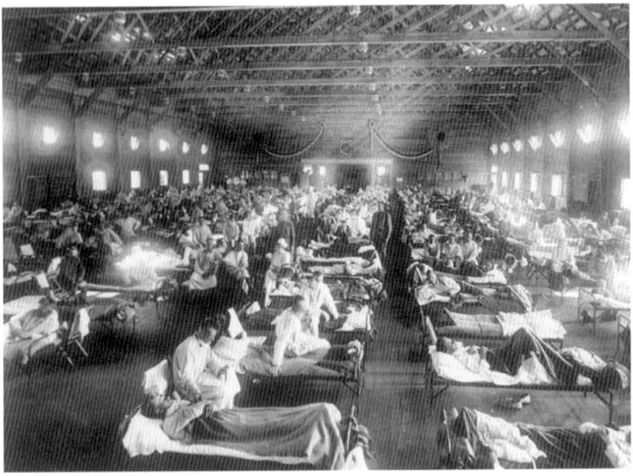

Figure 2. Emergency hospital during the Spanish flu pandemic in 1918, Camp Funston, Kansas, USA. Courtesy of the National Museum of Health and Medicine, Armed Forces Institute of Pathology, Washington, DC, USA (NCP1603).

Pandemics and epidemics

There are many different influenza A virus subtypes, differing in the nature of the HA and NA glycoproteins on their surface.[12,13] Sixteen HAs and nine NAs have been identified. Virus subtypes are distinguishable serologically, which means that antibodies to one subtype do not react with another subtype. Aquatic birds appear to serve as the reservoir of influenza A viruses; indeed, all virus subtypes presumably circulate in these wild birds. Only H1, H2 and H3 virus subtypes have been identified in humans, specifically the H1N1, H2N2 and H3N2 viruses corresponding to the three major pandemics of the last century. During recent outbreaks of avian influenza, there have been occasional transmissions of H5N1, H7N7 and H9N2 viruses to humans (see Chapter 3).

Antigenic shift

Influenza pandemics are the result of so-called antigenic shift of the virus (see Chapter 3).[13] This means that a virus with a new HA (and NA) is introduced into the human population. Since everyone will be immunologically naive to that new virus subtype, the infection may spread rapidly and cause high morbidity and mortality among the entire population, including young healthy people.[3,14] Three major influenza pandemics struck the globe in the 20th century.[3,4] The Spanish flu (caused by an H1N1 influenza A virus) occurred in 1918–19, after the First World War, and resulted in the death of an estimated 50 million people,[14] more than the death toll of the war itself (Figure 2). Other pandemics occurred in 1957 (Asian flu, H2N2) and 1968 (Hong Kong flu, H3N2), with a total number of deaths of approximately 2 million.[1]

The current information supports the concept that new pandemic influenza viruses are derived from avian virus reservoirs.[13,15–17] Avian influenza viruses may be directly transmitted to humans, which probably occurred in the case of the 1918 Spanish flu virus.[18,19] The possibility of direct transmission of an avian influenza virus to humans became evident for the first time during the H5N1 outbreak in Hong Kong in 1997.[20,21] In this incident, 18 people were infected with the chicken flu virus, six of whom died. Similarly, in 2003, an outbreak of fowl plague (H7N7 subtype) in the Netherlands resulted in the direct transmission of the virus to poultry workers and veterinarians, killing one person.[22] Furthermore, as of the end of 2003, continuing renewed outbreaks of highly pathogenic H5N1 avian influenza in Asia, and more recently also in Europe, Africa and the Middle East, have resulted in an increasing number of human infections with a very high case-fatality rate,[23-27] underscoring the disconcerting possibility that this virus may further adapt to the human host and thus cause a new deadly worldwide outbreak of influenza.

Antigenic shift may also be due to a genetic reassortment between an avian and a human influenza virus, with pigs possibly acting as an intermediate host. In this process, a human virus acquires a number of gene segments from an avian virus, at least including the gene segment encoding the HA, within a cell dually infected with two viruses from human and avian origin (see Chapter 3). It is well established that genetic reassortment has been the basis of the formation of the 1957 and 1968 pandemic viruses.[28]

Antigenic drift

In addition to the emergence of new human influenza virus subtypes with pandemic potential, established human virus subtypes undergo significant antigenic adaptation, referred to as antigenic drift (see Chapter 3).[13] Antigenic drift involves minor changes in the HA, NA and possibly also other viral antigens, that occur due to mutations in the viral genome, resulting in amino acid substitutions in antigenic sites. These changes may render the new strain different enough to at least partially avoid the immunity induced by previous strains. Thus, new influenza epidemics may arise. In the temperate regions, these epidemics generally occur in the winter months (from October to April in the northern hemisphere, and from April to October in the southern hemisphere). Antigenic drift of established human virus subtypes necessitates regular update of the composition of the annual influenza vaccine (see Chapter 8).

Clinical picture and impact of flu

Transmission and pathogenesis

After being inhaled by aerosol or droplet transmission, the influenza virus attacks the respiratory epithelium and commandeers the host cell's replication machinery to produce new progeny virus particles, which further infect other cells (see Chapters 2 and 5). "Innate" immune responses retard the initial spread of the virus in the respiratory tract and provide valuable time for the host to mount a specific "adaptive" immune response against the invading virus. B-lymphocytes produce antibodies that bind to and neutralize the virus on the mucosal surfaces, while T-lymphocytes dispose of the infected cells that have become factories for producing new virus particles (see Chapter 4).

The body's immune response to influenza produces a number of inflammatory cytokines, which are responsible for most of the symptoms of influenza. Except in rare circumstances, the virus remains limited to the respiratory epithelium (see Chapter 2). Thus, the major systemic symptoms such as fever, myalgia and malaise are caused by these circulating cytokines. Influenza is generally a self-limiting disease. However, in the complete absence of immunity to a specific subtype of influenza, such as with the emergence of a novel pandemic strain, even young people may die within 24–48 hours of the onset of symptoms, the overwhelming inflammatory response to the virus causing an acute respiratory distress-like syndrome. Inadequate immune responses as often occur in the very young and very old may lead to a primary viral pneumonia or secondary bacterial infection (see Chapter 5).

Impact of influenza epidemics and pandemics

Influenza epidemics have been recognized as a major cause of morbidity and increased mortality, especially in the very young, the very old, people with chronic cardiopulmonary conditions, pregnant women and immunocompromised individuals (see Chapter 6). Each year, influenza results in 3–5 million cases of severe illness and kills between 0.25 and 0.5 million people worldwide.[7,29] The WHO estimates that there are currently 1 billion people worldwide who are at high risk of suffering or dying from influenza and its complications.[29] Furthermore, as the elderly population increases, future influenza epidemics will be associated with ever-increasing hospitalization rates and excess mortality unless adequate prophylactic measures are taken. In developed countries, about 100 people per million population die annually from influenza.[7,30] Most of these deaths (95%) occur in those over 60 years of age (12%, 29% and 54% in people aged 60–69, 70–79 and >80 years respectively).[31]

As indicated above, the three major pandemic outbreaks of influenza in the 20th century have caused many millions of deaths, and future pandemics are also expected to have a devastating impact on society.[32] Yet, it has been estimated that the cumulative mortality of annual influenza epidemics in the last century exceeded that of the three major pandemic outbreaks.[7,33] This underscores the fact that not only influenza pandemics but also the recurrent annual outbreaks of flu remain a major health threat worldwide. Thus, public-health policy needs to address the interrelated issues of optimizing both the effectiveness of annual vaccination programmes and the preparedness for future pandemic outbreaks of influenza. Health-care professionals play a key role in this regard by participating in influenza surveillance and by providing appropriate preventive and treatment regimens for their patients.

Prevention and control

Vaccination

Vaccination remains the cornerstone of influenza prevention (see Chapter 8). Inactivated influenza vaccines have been in use for the past 60 years (Figure 3). These vaccines have an excellent safety record and have proven to be efficacious in preventing influenza and reducing complications and deaths. The WHO recommends annual vaccination of people in at-risk groups.[8,9] The primary target groups for annual vaccination include the elderly, nursing-home residents, patients with chronic respiratory

Figure 3. Influenza vaccination in the 1950s. Photograph courtesy of Solvay Pharmaceuticals, Weesp, the Netherlands.

or cardiovascular disease, diabetes or renal dysfunction, as well as immunocompromised individuals. In addition, it is also recommended that health-care workers and family members of people in high-risk groups be vaccinated.

The clinical effectiveness and cost-effectiveness of inactivated influenza vaccines have been clearly demonstrated.[34] For example, vaccination of community-dwelling senior citizens reduces hospitalization rates for pneumonia or other respiratory conditions by >30% and death from all causes by 50%. For nursing-home residents the benefits of influenza vaccination are even more striking, with reductions of hospitalization rates or death from all causes of 48% and 68%, respectively (see Chapter 8).

However, despite the established effectiveness of immunization, the national and international response to influenza prevention measures is often inadequate (see Chapter 8). In many countries, the established benefits of vaccination have not been translated into effective immunization programmes. Only 50 countries have policies for influenza immunization and often only 10–20% of people in high-risk groups, particularly those below the age of 65, are protected.[10,29] While we do have the means to prevent the serious consequences of influenza, the implementation of preventive measures is still suboptimal, and as a consequence, currently many at-risk individuals remain unprotected and vulnerable to infection and death.[11,29]

"The relatively low uptake of influenza vaccines in most industrialized countries implies that significant proportions of the groups at risk of complications from influenza are not vaccinated. WHO strongly emphasizes the importance of raising the public consciousness of influenza and its complications as well as of the beneficial effects of influenza vaccination."

***World Health Organization* (Ref. 9; Appendix 3)**

How can vaccine uptake be further encouraged?

As indicated above, the WHO recommends annual immunization of at-risk persons as the best and most cost-effective strategy for reducing influenza-related morbidity and mortality.8,9,29 However, there is often a gap between policy, on the one hand, and practice and uptake by the population, on the other, especially when interventions depend on people to change their behaviour to lower disease risk. Public-health policy is most effectively translated into improved health outcomes when social and individual contexts are clearly understood and relevant communities are engaged in the process.

Health-care workers, because of their direct relationship with the patient, are in an optimal position to promote preventive measures and explain the facts about influenza, its complications and the safety and efficacy of influenza vaccination. Accordingly, the WHO encourages initiatives to raise the awareness, particularly among health-care professionals, about the impact of influenza and the benefits of influenza vaccination.9 On the basis of the available evidence, offering vaccination to at-risk patients is no less than an ethical obligation.11 Indeed, by ensuring that all people in target groups are vaccinated, according to the recommendations of the WHO8,9 and many national health authorities, the health-care worker can make a significant contribution to public health. If, ultimately, health-care workers would decide to vaccinate the majority of their at-risk patients, this would go a long way towards ameliorating the burden of future influenza epidemics and pandemics.

Antivirals

While vaccination is the method of choice for influenza prophylaxis, under specific conditions where the individual has not been or cannot be vaccinated, or is not fully protected by vaccination, the use of antiviral drugs should be considered for treatment or prevention of influenza infection (see Chapter 7). In this regard, the physician needs to know whether influenza is circulating in the community and how to diagnose influenza illness for the purpose of appropriate prescription of influenza antiviral drugs.

Antiviral drugs approved for treament or prophylaxis of influenza include the M2 channel inhibitors amantidine and rimantidine, and the neuraminidase inhibitors zanamivir and oseltamivir. The application of these drugs to clinical practice may be limited by issues of efficacy with respect to the type of influenza virus (influenza A vs B), drug resistance, adverse effects and cost. In the case of the M2 ion channel blockers, these drugs are only effective against influenza A, the induction of drug resistance is well documented and side effects can be significant, especially in older people. While these issues have not been apparent with the more recent neuraminidase inhibitors, high cost may be prohibitive for broad use of these drugs, while for the inhaled zanamivir there is a caution against its use in cases of airway hypersensitivity. These issues not only limit the use of antivirals for individual treatment, but also present major challenges to their potential use in future pandemics (see Chapter 7).

Is the world prepared for the next pandemic?

Since the 1997 H5N1 avian flu outbreak in Hong Kong, the need for intensified measures to control future influenza pandemics has been widely recognized. In 2003, a resolution of the World Health Assembly urged member states to increase their efforts in the area of influenza prevention.[29] In November 2005, during a large international meeting with participation of the WHO, the World Bank, the UN co-ordinator for pandemic preparedness and delegates from over 100 countries,[35] there was broad consensus that the threat of a pandemic remains. While the mass culling of poultry in Hong Kong in 1997 (Figure 4) may have averted a pandemic, in the meantime not only has the H5N1 virus reappeared, but it has now spread throughout many countries in Asia and parts of Europe, Africa and the Middle East, probably carried by infected migratory water fowl. During these ongoing H5N1 outbreaks, there have

Figure 4. Culling of chickens during the 1997 outbreak of H5N1 avian flu in Hong Kong. Copyright © Reuters/CORBIS.

been 179 confirmed human infections (March 1, 2006), with a case-fatality rate of over 50%.[27] Clearly, it is only a matter of time before another pandemic influenza virus will emerge. This will inevitably result in the death of millions of people unless preventive measures are adequately instituted.

However, despite the imminent pandemic threat, most countries and the world at large are still not sufficiently prepared to contain a new global influenza outbreak. For example, in the current situation, no country will have vaccines at the start of the pandemic and production of significant quantities of vaccine will take many months after the first detection of the new pandemic virus[36] (see also Chapter 9). There are also concerns about the efficacy of antivirals for both treatment and prophylaxis in the setting of a pandemic, which may thwart antiviral stockpiling strategies[37] (see also Chapter 7). Therefore, there remains a need for more preparatory action.

Fortunately, many countries now do have pandemic preparedness plans in place which describe how these countries will deal with the emergency and what steps will be taken to contain a potential future pandemic as much as possible. The single most important aspect of further improved national as well as international pandemic preparedness involves the stimulation of annual influenza vaccine uptake among target groups (see Chapter 8) along with the development of more effective vaccine formulations including procedures to generate these vaccines as rapidly as possible once the pandemic emerges (see Chapter 9). The availability of more effective vaccines will permit individual countries and the global community at large to distribute the scarce quantities of vaccine antigen in an equitable fashion, such that not only the people in the wealthy countries, but also those in the Third World, will be protected as much as possible.

References

1. World Health Organization. Avian influenza: assessing the pandemic threat. 2005. (www.who.int/csr/disease/influenza/WHO_CDS_2005_29/en/)

2. Patterson KD. Pandemic and epidemic influenza, 1830–1848. *Soc Sci Med* 1985; **21**: 571–580.

3. Ghendon Y. Introduction to pandemic influenza through history. *Eur J Epidemiol* 1994; **10**: 451–453.

4. Beveridge WI. The chronicle of influenza epidemics. *Hist Phil Life Sci* 1991; **13**: 223–234.

5. Shope RE. Swine influenza. III. Filtration experiments and etiology. *J Exp Med* 1931; **54**: 373–380.

6. Smith W, Andrewes C, Laidlaw P. A virus obtained from influenza patients. *Lancet* 1933; **2**: 66–68.

7. Nicholson KG, Wood JM, Zambon M. Influenza. *Lancet* 2003; **362**: 1733–1745.

8. World Health Organization. Influenza vaccines. *Wkly Epidemiol Rec* 2000; **75**: 281–288. (www.who.int/docstore/wer/pdf/2000/wer7535.pdf)

9. World Health Organization. Influenza vaccines. *Wkly Epidemiol Rec* 2002; **77**: 230–239. (www.who.int/docstore/wer/pdf/2002/wer7728.pdf)

10. The Macroepidemiology of Influenza Vaccination (MIV) Study Group. The macroepidemiology of influenza vaccination in 56 countries, 1997–2003. *Vaccine* 2005; **23**: 5133–5143.

11. Palache A. Influenza vaccines. A reappraisal of their use. *Drugs* 1997; **54**: 841–856.

12. Lamb RA, Krug RM. *Orthomyxoviridae*: the viruses and their replication. In: Knipe DM, Howley PM, Griffin DE *et al.*, editors. *Fields Virology*, 4th edn. Lippincott Williams & Wilkins, 2001; pp. 1487–1531.

13. Wright PF, Webster RG. Orthomyxoviruses. In: Knipe DM, Howley PM, Griffin DE *et al.*, editors. *Fields Virology*, 4th edn. Lippincott Williams & Wilkins, 2001; pp. 1533–1579.

14. Johnson NP, Mueller J. Updating the accounts: global mortality of the 1918–1920 "Spanish" influenza pandemic. *Bull Hist Med* 2002; **76**: 105–115.

15. Subbarao K, Katz J. Avian influenza viruses infecting humans. *Cell Molec Life Sci* 2000; **57**: 1770–1784.

16. Webster RG. Influenza virus: transmission between species and relevance to emergence of the next human pandemic. *Arch Virol Suppl* 1997; **13**: 105–113.

17. Zambon M. The pathogenesis of influenza in humans. *Rev Med Virol* 2001; **11**: 227–241.

18. Taubenberger JK, Reid AH, Lourens RM *et al.* Characterization of the 1918 influenza virus polymerase genes. *Nature* 2005; **437**: 889–893.

19. Taubenberger JK, Morens DM. 1918 influenza: the mother of all pandemics. *Emerg Inf Dis* 2006; **12**: 15–22.

20. Claas EC, Osterhaus AD *et al.* Human influenza A H5N1 virus related to a highly pathogenic avian influenza virus. *Lancet* 1998; **351**: 472–477.

21. Subbarao K, Klimov A, Katz J *et al.* Characterization of an avian influenza A (H5N1) virus isolated from a child with a fatal respiratory illness. *Science* 1998; **279**: 393–396.

22. Fouchier RAM, Schneeberger PM, Rozendaal FW *et al.* Avian influenza A virus (H7N7) associated with human

conjunctivitis and a fatal case of acute respiratory distress syndrome. *Proc Natl Acad Sci USA* 2004; **101**: 1356–1361.

23. Peiris JS, Yu WC, Leung CW *et al.* Re-emergence of fatal human influenza A subtype H5N1 disease. *Lancet* 2004; **363**: 617–619.

24. Apisarnthanarak A, Kitphati R, Thongphubeth K *et al.* Atypical avian influenza (H5N1). *Emerg Infect Dis* 2004; **10**: 1321–1324.

25. de Jong MD, Bach VC, Phan TQ *et al.* Fatal avian influenza A (H5N1) in a child presenting with diarrhea followed by coma. *New Engl J Med* 2005; **352**: 686–691.

26. Tran TH, Nguyen TL, Nguyen TD *et al.* Avian influenza A (H5N1) in 10 patients in Vietnam. *New Engl J Med* 2004; **350**: 1179–1188.

27. World Health Organization. Cumulative number of confirmed human cases of avian influenza A/(H5N1) reported to WHO. (www.who.int/csr/disease/avian_influenza/country/en/index.html).

28. Ito T, Couceiro JN, Kelm S *et al.* Molecular basis for the generation in pigs of influenza A viruses with pandemic potential. *J Virol* 1998; **72**: 7367–7373.

29. Influenza: Report by the WHO Secretariat for 111th session of the WHO Executive Board, January 2003. Document EB111/10, WHO, November 2002. (www.who.int/gb/ebwha/pdf_files/EB111/eeb11110.pdf)

30. Barker WH, Mullooly JP. Impact of epidemic type A influenza in a defined adult population. *Am J Epidemiol* 1980; **112**: 798–811.

31. Sprenger MJ, Mulder PG, Beyer WE *et al.* Impact of influenza on mortality in relation to age and underlying disease, 1967–1989. *Int J Epidemiol* 1993; **22**: 334–340.

32. Meltzer MI, Cox NJ, Fukuda K. The economic impact of pandemic influenza in the United States: priorities for intervention. *Emerg Infect Dis* 1999; **5**: 659–671.

33. Simonson L, Clarke MJ, Williamson GD *et al.* The impact of influenza epidemics on mortality: introducing a severity index. *Am J Public Health* 1997; **87**: 1944–1950.

34. Nichol KL. The efficacy, effectiveness and cost-effectiveness of inactivated influenza virus vaccines. *Vaccine* 2003; **21**: 1769–1775.

35. World Health Organization. Avian influenza and human pandemic influenza. Summary report of meeting held in Geneva, 7–9 November 2005. (www.who.int/mediacentre/events/2005/avian_influenza/summary_report_Nov_2005_meeting.pdf)

36. Luke CJ, Subbarao K. Vaccines for pandemic influenza. *Emerg Inf Dis* 2006; **12**: 66–72.

37. Monto A. Vaccine and antiviral drugs in pandemic preparedness. *Emerg Inf Dis* 2006; **12**: 55–60.

2. The Influenza Virus: Structure and Replication

Influenza viruses are enveloped RNA viruses, belonging to the family *Orthomyxoviridae*.[1] There are three influenza virus genera, or virus types, within this family, influenza A, B and C, distinguishable on the basis of antigenic differences between their matrix and nucleoproteins (M and NP). Influenza A, B and C viruses also differ with respect to host range, variability of the surface glycoproteins, genome organization and morphology.[1] The influenza A viruses are responsible for pandemic outbreaks of influenza and for most of the well-known annual flu epidemics.[2] Therefore, the discussion here will be limited primarily to influenza A viruses, only referring to influenza B where appropriate. Influenza C virus, which is substantially different from the A and B viruses, is of little importance for human influenza infections, causing only a mild common-cold-like disease; it will not be further discussed in this book.

The A and B viruses contain two major envelope glycoproteins, haemagglutinin (HA) and neuraminidase (NA).[1] An important feature of influenza viruses is their segmented genome, containing eight independent RNA strands of negative polarity. The RNA segments are packaged in the viral core. The core is surrounded by a lipid membrane, or "envelope", derived from the plasma membrane of the infected host cell during the process of budding of progeny virus from the cell's surface.[1]

Human-to-human transmission of influenza occurs through aerosols or droplets, spread into the environment by a sneezing or coughing infected individual.[3] The virus attacks primarily epithelial cells of the upper and lower respiratory tract.[2] Infection occurs by binding of the viral HA to sialic acid receptors on the target cell surface and subsequent fusion of the viral envelope with the host cell membrane.[1] It is through this fusion process that the viral

RNA gains access to the cytosol of the host cell. The RNA then enters the nucleus of the cell, where it is replicated. Viral proteins are being synthesized in the cytosol, ultimately resulting in production of many new virus particles. During the production of progeny virus, the host cell's own protein synthesis is effectively shut down. Finally, having produced many thousands of new virus particles, the cell lyses and dies as a result of the infection.[1]

Key Messages

- There are three types of influenza virus: A, B and C. Influenza A viruses are responsible for all pandemic and most epidemic outbreaks, influenza B viruses also cause human disease, but influenza C virus is of little clinical importance.
- Influenza A and B viruses have two major envelope glycoproteins, haemagglutinin (HA) and neuraminidase (NA). HA is responsible for infectious entry of the virus into cells; it is also the virus' most important surface antigen, against which virus-neutralizing antibodies are directed.
- Influenza virus enters cells through receptor-mediated endocytosis and low-pH-induced fusion from within acidic endosomes.
- Cleavage into two subunits is essential for HA to mediate entry of influenza virus into cells. The HA1 subunit is involved in receptor binding, HA2 has membrane fusion activity.
- The receptor for HA is sialic acid bound to glycolipids or glycoproteins on the host cell surface. Avian and human influenza A viruses have a different receptor specificity; pigs have receptors for both avian and human viruses.
- NA cleaves the sialic acid receptor, thus releasing progeny virus from the infected cell surface. It is the target for the antiviral drugs zanamivir and oseltamivir, which are sialic acid analogues and inhibit release of progeny virus from infected cells.
- Influenza viruses have a segmented RNA genome of negative polarity. Mixed infection of a single cell with two different influenza viruses may lead to genetic reassortment between the two viruses.

Classification and nomenclature of influenza viruses

Influenza A viruses are divided into subtypes, based on the nature of their surface glycoproteins, HA and NA.[1] There are 16 different HAs and nine NAs which are distinguishable serologically, i.e. antibodies to one virus subtype do not react with another. All virus subtypes appear to circulate in aquatic birds.[4] Only some of these subtypes have been identified in humans (Table 1), specifically the H1N1, H2N2 and H3N2 viruses, corresponding to the three major pandemics of the last century. During recent outbreaks of highly pathogenic avian influenza (HPAI), there have been occasional transmissions of H5N1, H7N7 and H9N2 viruses to humans (see Chapter 3). Influenza A viruses are known to also infect a variety of other mammals, including non-human primates, pigs, horses, cats, seals, whales and mink (Table 1).[1,2,4] There are no influenza B virus subtypes. The B virus primarily infects humans, although it has also been isolated from seals.[5]

The current nomenclature system for influenza A viruses includes the host of origin, geographic location of first isolation, strain number and year of isolation.[6] The HA and NA subtype of influenza A viruses is specified in parentheses, e.g. A/Swine/Iowa/15/30 (H1N1). By convention, the host of origin of human strains is omitted, e.g. A/Puerto Rico/8/34 (H1N1). Since there are no subtypes of influenza B virus, no parenthetical specification is given in this case and, as the B viruses primarily infect humans, the host of origin is not mentioned in the influenza B virus nomenclature, e.g. B/Yamagata/16/88.

Influenza virus structure

Influenza viruses are roughly spherical, although somewhat pleomorphic, particles, ranging from 80 to 120 nm in diameter.[1,7] Figure 5 presents a model of the overall structure of the influenza virus. A characteristic feature of influenza virus particles is their external layer of

Natural hosts of influenza A viruses			
Haemagglutinin		**Neuraminidase**	
Subtype	**Predominant hosts**	**Subtype**	**Predominant hosts**
H1	Human, pig, birds	N1	Human, pig, birds
H2	Human, pig, birds	N2	Human, pig, birds
H3	Birds, human, pig, horse	N3	Birds
H4	Birds	N4	Birds
H5	Birds, (human)	N5	Birds
H6	Birds	N6	Birds
H7	Birds, horse, (human)	N7	Horse, birds
H8	Birds	N8	Horse, birds
H9	Birds, (human)	N9	Birds
H10	Birds		
H11	Birds		
H12	Birds		
H13	Birds		
H14	Birds		
H15	Birds		
H16	Birds		

Table 1. Natural hosts of influenza A viruses. The table indicates the subtypes of haemagglutinin (HA) and neuraminidase (NA), and the hosts in which they have been identified. Adapted from Lamb RA, Krug RM. *Orthomyxoviridae*: the viruses and their replication. In: Knipe DM, Howley PM, Griffin DE *et al*., editors. *Fields Virology*, 4th edn. Lippincott Williams & Wilkins, 2001; pp. 1487–1531[1] with permission from Lippincott Williams & Wilkins.

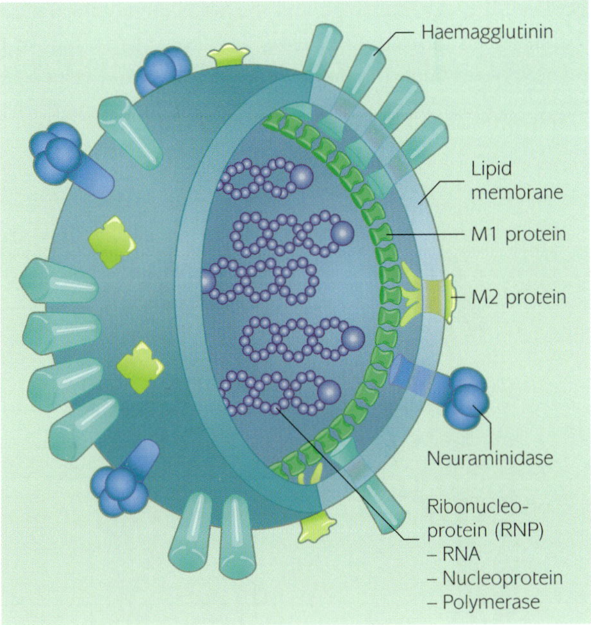

Figure 5. Model of influenza virus.

approximately 500 spike-like projections. These spikes represent the envelope glycoproteins HA (which has a rod-like shape) and NA (which is mushroom-shaped).[7] The HA spike is a trimer, consisting of three individual HA monomers,[8] while the NA spike is a tetramer.[9,10] HA is about four times more abundant than NA.

The viral envelope proteins

The major envelope glycoprotein HA is synthesized in the infected cell as a single polypeptide chain (HA0) with a length of approximately 560 amino acid residues, which is subsequently cleaved into two subunits, HA1 and HA2.[1,8] These subunits remain covalently linked to one another through disulphide bonds. Cleavage of HA0 is essential for the molecule to be able to mediate membrane fusion between the viral envelope and the host cell membrane, as discussed in more detail below. HA belongs to the first proteins for which the entire three-dimensional structure has been elucidated. Treatment of whole virions with the

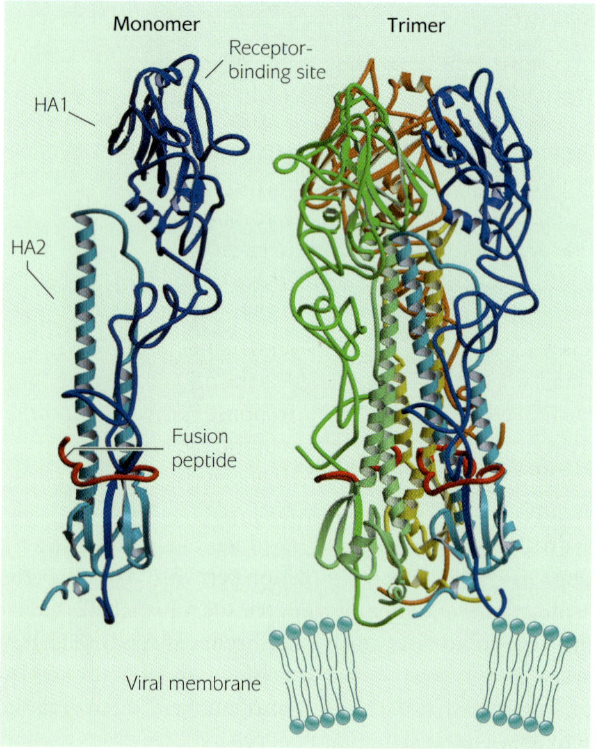

Figure 6. The three-dimensional structure of the influenza haemagglutinin (HA). The HA monomer (left) and trimer (right) are shown. In the monomer, the globular HA1 subunit is shown in dark blue, the HA2 subunit in light blue, with the "fusion peptide" in red. The receptor-binding site of HA1 is located at the tip of the molecule. This figure was produced by André van Eerde (University of Groningen), using MOLSCRIPT, on the basis of the co-ordinate file from the Protein Data Bank, code 3HMG. Weis WI *et al*. Refinement of the influenza virus hemagglutinin by simulated annealing. *J Molec Biol* 1990; **212**: 737–761.

enzyme bromelain, which clips the polypeptide chain just above the viral membrane, releases a water-soluble fragment of the HA spike, commonly referred to as BHA. The ectodomain of HA of A/Aichi/2/68 (H3N2) virus, related to the Hong Kong pandemic virus of 1968, has thus been crystallized and subjected to X-ray analysis.[8,11] Figure 6 presents a representation of the 3D structure of HA based on this pioneering X-ray crystallographic structure determination.

The HA spike protrudes approximately 13.5 nm from the viral surface.[11,12] HA1 and HA2 appear in the structure of the spike as distinct subunits. HA1, the globular domain at the distal end of the spike, is responsible for binding of the virus to its cellular sialic acid receptor, the receptor-binding pocket being located close to the very tip of the molecule (Figure 6).[12] HA1 also contains the major antigenic epitopes of the molecule (Figure 7).[13] As discussed in more detail in Chapter 4, HA is the primary viral antigen to which the host's antibody response is directed and the only antigen inducing a virus-neutralizing response.

HA2 forms the fibrous stem of the viral spike. The N-terminus of HA2 contains a conserved stretch of 20, mostly hydrophobic, amino acid residues. This sequence is generally referred to as the "fusion peptide"; it triggers the membrane fusion process between the viral envelope and the host cell membrane,[11,12] as discussed in more detail below.

The second envelope glycoprotein NA has enzymatic activity, cleaving sialic acid residues from glycoproteins or glycolipids.[9] Since sialic acid functions as a receptor for attachment of influenza virions, the neuraminidase activity of NA, cleaving such receptors,[14] mediates the release of newly formed virus particles from the surface of infected cells.[15] NA is the target for the antiviral drugs oseltamivir (Tamiflu®) and zanamivir (Relenza®) (see Chapter 7). These drugs are sialic acid analogues,[16] which inhibit the enzymatic activity of NA, thus slowing down the release of progeny virus from infected cells.

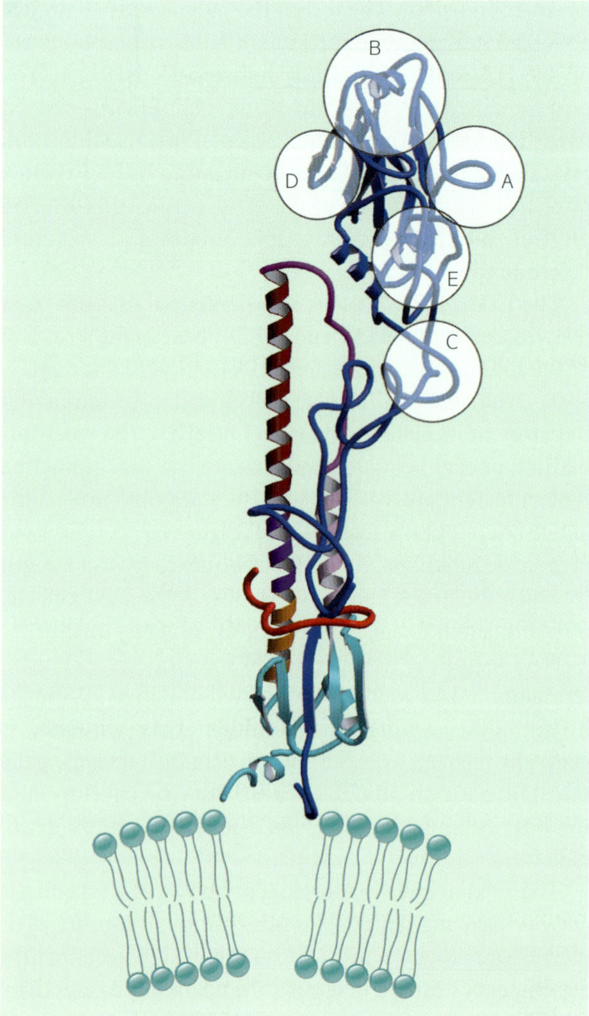

Figure 7. The location of the five major antigenic epitopes, A–E, on the HA1 subunit of the influenza virus haemagglutinin.[13] In the intact HA trimer, epitope D is not exposed and thus may not be involved in antibody induction. This figure was produced as described in the caption to Figure 6.

The influenza A virus envelope contains a small number of copies of a third integral membrane protein, M2, which forms a tetramer with ion channel activity.[17–19] M2 is involved in the infection process by modulating the pH within virions, weakening the interaction between the viral ribonucleoproteins (RNPs) and the M1 protein. M2 is the target for the anti-influenza drugs amantadine and rimantadine.[20] Influenza B viruses also contain a similarly limited number of copies of the integral membrane protein NB.[17] This protein may well be a functional homologue of the M2 protein of the A viruses, but it is not inhibited by amantadine and rimantadine. Thus, these antiviral drugs are not effective against influenza B (see also Chapter 7).

The viral core

The influenza A or B virus genome consists of eight segments of negative-sense single-stranded RNA.[1] Each RNA segment is associated with multiple copies of NP and with the viral transcriptase consisting of RNA polymerase components PB1, PB2 and PA, thus forming the RNP complex.[21] The RNPs are surrounded by a layer of the matrix protein, M1. With approximately 3000 copies per virion, M1 is the most abundant structural protein of influenza virus.[1]

RNA segments 1–6 of influenza A viruses encode a single protein each.[1] For example, segment 4 encodes the HA, segment 5 NP and segment 6 the NA protein. Segment 7 encodes two proteins, M1 and M2, with overlapping reading frames. Likewise, segment 8 encodes the non-structural proteins NS1 and NS2, again with superimposed reading frames. Even though NS2 was originally thought to be absent from virus particles (hence the name "non-structural"), it has subsequently been found in low copy numbers in virions. NS1 is not present in virions, but it is abundant in infected cells. Table 2 presents a survey of the RNA segments and the corresponding gene products of influenza A viruses.

Influenza A virus RNA segments and the proteins they encode		
RNA segment (no. of nucleotides)	**Gene product (no. of amino acids)**	**Molecules per virion**
1 (2341)	Polymerase PB2 (759)	30–60
2 (2341)	Polymerase PB1 (757)	30–60
3 (2233)	Polymerase PA (716)	30–60
4 (1778)	Haemagglutinin (566)	500
5 (1565)	Nucleoprotein (498)	1000
6 (1413)	Neuraminidase (454)	100
7 (1027)	Matrix protein M1 (252) Matrix protein M2 (97)	3000 20–60
8 (890)	Non-structural proteins • NS1 (230) • NS2 (121)	 – 130–200

Table 2. Influenza A virus RNA segments and the proteins they encode. Influenza A viruses have eight gene segments encoding 10 different proteins, segments 7 and 8 encoding two proteins each. Adapted from Lamb RA, Krug RM. *Orthomyxoviridae*: the viruses and their replication. In: Knipe DM, Howley PM, Griffin DE *et al.*, editors. *Fields Virology*, 4th edn. Lippincott Williams & Wilkins, 2001; pp. 1487–1531[1] with permission from Lippincott Williams & Wilkins.

Virus replication

Receptor binding and cell entry

In humans, the primary targets for influenza viruses are epithelial cells in the upper and lower respiratory tract. As indicated above, the viral HA binds to sialic acid residues on glycoproteins or glycolipids on the cell surface.[9] The fine specificity of HA's receptor binding depends on the nature of the glycosidic linkage between the terminal sialic acid and the penultimate galactose residue on the receptor.[22] Human influenza viruses preferentially bind to sialic acids attached to galactose in an α2,6 configuration, whereas avian viruses have a preference for sialic acids attached to galactose in an α2,3 linkage.[23] This difference is thought to be the basis for the very inefficient transmission of avian influenza viruses to humans. Pigs, on the other hand, have receptors with either type of linkage between sialic acid and galactose, and thus are readily susceptible to infection with both human and avian viruses.[23] Co-infection of pigs with different influenza viruses is considered one mechanism by which new influenza viruses with pandemic potential may arise (see also Chapter 3).

Receptor binding initiates uptake of the virus through so-called receptor-mediated endocytosis. In this process, virus particles are engulfed by the host cell plasma membrane (Figure 8). The vesicles thus formed subsequently fuse with intracellular compartments called endosomes, as a result of which the virus is delivered to the endosomal lumen.[24–26] Receptor-mediated endocytosis is not specific for uptake of viruses. In fact, it is a general mechanism by which cells internalize macromolecular complexes. Substances taken up by endocytosis, having traversed the endosomal cell compartment, generally end up in lysosomes, where they are degraded by hydrolytic enzymes. The influenza virus genome, however, escapes degradation; through fusion of the viral envelope with the endosomal membrane, it gains access to the cell cytosol (Figure 8).

Membrane fusion and uncoating of the viral core

It is the low pH inside the endosomes (pH 5–6), maintained by proton pumps within the endosomal membrane, that triggers the fusion reaction between the viral envelope and the endosomal membrane.[24–27] This is a key step in the viral infection mechanism. At low pH, a major conformational change in the HA spike is induced. This conformational

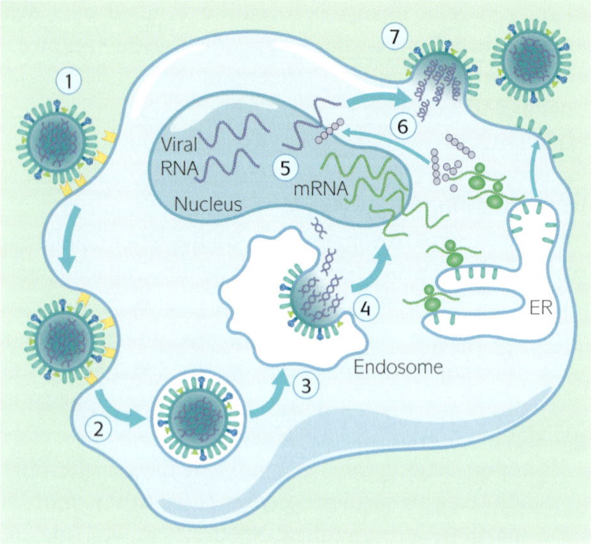

Figure 8. Life cycle of influenza virus. (1) Binding of the virus to a sialic acid-containing receptor. (2) Engulfment of the virus by the cell plasma membrane and formation of an endocytic vesicle. (3) Delivery of the virus to the endosomal cell compartment. (4) Fusion of the viral membrane with the membrane of the endosome, induced by the mildly acidic pH in the endosomal lumen. (5) Delivery of viral RNA to the nucleus, synthesis of messenger RNA (mRNA) and viral RNA replication. (6) Synthesis of viral protein components in the cell cytosol (internal proteins) and endoplasmic reticulum (ER) (membrane proteins). (7) Assembly and budding of progeny viruses. Adapted with kind permission of Linda Stannard.

change results in movement of the fusion peptide sequences of HA2, previously buried within the stem of the HA trimer, to the distal tip of the HA spike, allowing their insertion into the target membrane (Figure 9).[28,29] Subsequently, a complex process of bending of the trimer takes place promoted by the formation of a stable coiled coil structure consisting of heptad repeat regions close to the fusion peptide and the transmembrane anchor of

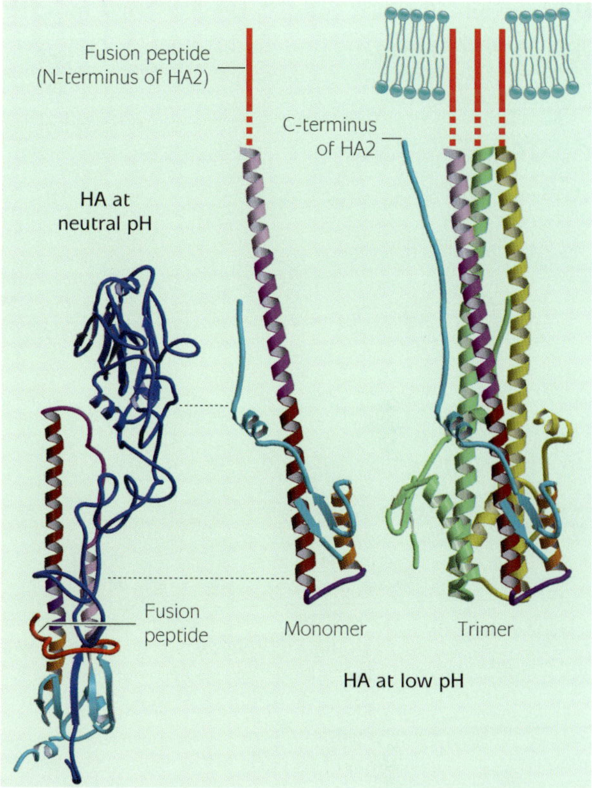

Figure 9. Conformational changes in the viral HA occurring at the pH of membrane fusion. The figure shows an HA monomer at neutral pH (left) and the low-pH form of HA2 both as a monomer and as a trimer (right). The long helices are represented in equivalent positions. As a result of the acid-induced conformational change in the molecule, the HA2 fusion peptides move upwards by about 10 nm to the tip of the trimer, such that they may insert into the endosomal membrane. Then the protein folds and induces membrane fusion, the fusion peptide and the C-terminal membrane anchor ultimately ending up in the same fused membrane. This figure was produced using MOLSCRIPT, as described in the caption to Figure 6, on the basis of co-ordinate files from the Protein Data Bank, codes 3HMG and 1HTM. Bullough PA *et al*. Structure of influenza haemagglutinin at the pH of membrane fusion. *Nature* 1994; **371**: 37–43.[29]

HA2.[30] Thus, the two ends of HA2 inserted into the apposed membranes are brought together. This triggers

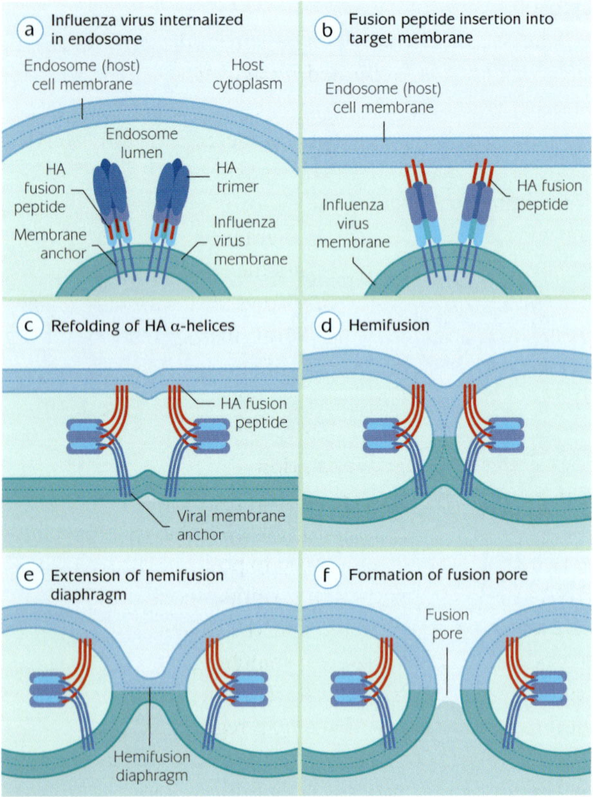

Figure 10. Hypothetical mechanism of HA-mediated fusion between the influenza virus membrane and the endosomal membrane, involving the formation of a hemifusion diaphragm. The key step in the fusion process is the relocation of the fusion peptides of HA2 to the tip of the HA trimer such that they can penetrate the endosomal membrane (b); this step is triggered by the low pH in the endosome. Adapted from Cross KJ *et al*. Mechanisms of cell entry by influenza virus. *Exp Rev Molec Med*, 2001, Vol 6. (www-ermm.cbcu.cam.ac.uk/01003453h.htm) with permission from Cambridge University Press.

the merging of the two membranes, involving the formation of a distinct hemifusion intermediate and the subsequent formation of a fusion pore (Figure 10) through which the viral genetic material gains direct access to the cytosol of the cell (Figure 8).

The release of the viral RNPs from the endosome into the cytoplasm is facilitated by acidification of the viral interior prior to the fusion step. This acidification is mediated by the M2 proton channel in the viral envelope mentioned above.[17,20,31] After exposure of the virus to pH 5–6 within the lumen of the endosome, protons flow into the viral interior, weakening interaction of the M1 protein layer with the viral envelope and the RNPs. Blocking of the M2 channel with amantadine slows the dissociation of M1 from the RNPs and the viral membrane, inhibiting subsequent steps in the viral life cycle (see also Chapter 7).[17,20,32]

RNA replication and translation

The RNP complexes released into the host cell cytosol are transported to the nucleus (Figure 8). Here, the negative-sense viral RNAs are transcribed to positive-sense messenger RNAs (mRNAs) by the transcriptase (consisting of PB1, PB2 and PA) carried with the RNPs.[1] The transcriptase, in a process referred to as "cap snatching", steals short cap regions from cellular mRNAs as primers for initiation of viral mRNA synthesis. These cap regions are required for efficient binding of ribosomes to the mRNA. Cap snatching by the viral transcriptase thus inhibits the synthesis of cellular proteins in favour of production of viral components. The mRNAs are transported back to the cell cytosol and translated to protein.

The negative-sense viral RNAs also serve as templates for production of exact positive-sense RNA copies (cRNA), which in turn direct the synthesis of multiple new copies of negative-sense viral RNAs. These genomic segments are transported back to the cell's cytosol for assembly of new virus particles.

Synthesis of the viral envelope proteins HA, NA and M2 starts in the cytosol, but already during synthesis, the growing polypeptide chains are transported to the endoplasmic reticulum where the proteins are glycosylated and folded into trimers and tetramers (Figure 8).[33,34] Subsequently, the proteins are transported through the Golgi apparatus and the trans-Golgi network to the plasma membrane of the cell. Along this pathway, several modifications are introduced, including the formation of disulphide linkages and modification of the oligosaccharide side chains. Since the pH inside the trans-Golgi network is mildly acidic, a premature fusion-activating conformational change in HA would likely be induced, had the virus not developed a protection mechanism against this. As it happens, the M2 protein, which is abundantly expressed in infected cells, transiently neutralizes the pH within the trans-Golgi network, such that HA may transit safely to the cell surface.[17,20] This represents a second important function of the M2 protein, besides its role in viral entry and uncoating discussed above.

Synthesis and folding of viral core proteins occur entirely in the cytosol. NP and the RNA polymerase components interact with newly synthesized viral RNA to form RNPs. The M1 protein starts to interact with the C-terminal domains of HA and NA on the cell plasma membrane, forming patches with a high density of HA and NA, excluding cellular plasma membrane proteins. Subsequently, newly formed RNPs interact actively with the M1 lining at these patches (Figure 8). This interaction also prevents re-entry of RNPs into the nucleus.

Assembly and release of new viral particles

After attachment of RNPs to M1 on the inner half of the cell plasma membranes, in an intriguing process of budding, new virus particles are assembled (Figure 8). In polarized epithelial cells, this process occurs exclusively on the apical side of the cells, HA and NA being sorted to this ("exterior") surface of the cells' plasma membrane.[35] As a result, progeny virus is released back to the airways and not to the systemic circulation.

It has long been thought that the packaging of RNPs into new virions occurs in a random fashion. Thus, many non-infectious virus particles would be formed with an incomplete set of RNA segments, whereas just a minority of particles would contain the full complement of the proper eight RNA segments. More recent observations indicate, however, that packaging of RNPs is not a random process, but rather favours the formation of infectious virus particles with all eight RNA segments required for efficient infection.[21,36]

Upon co-infection of a cell with two influenza viruses from different origins (e.g. avian and human), the virus assembly process may result in mixing of the RNA segments from the two viruses and formation of a new virus with an altered genetic make-up. This mixing of gene segments, referred to as genetic reassortment, is one mechanism by which new influenza viruses with pandemic potential may arise (see also Chapter 3).

After budding, the new virions are still attached to the cell surface through interaction of the HA with sialic acid residues on cellular glycoproteins or glycolipids. It is at this point that the viral NA cleaves the sialic acid, thus releasing the virions from the host cell's surface,[15] allowing them to spread further throughout the respiratory tract. Antibodies directed against the NA or neuraminidase inhibitors block NA's neuraminidase activity, thus preventing release and spreading of the new progeny viruses (see also Chapter 7). This also explains why antibodies against NA, as well as neuraminidase inhibitors, do not neutralize virus infection, but rather aid in ameliorating the infection process.

The entire process of viral infection seriously disrupts the normal physiology of the cell. This, in the case of influenza, as with many other acute lytic viral infections, eventually leads to cell death. Cell death results in desquamation of the respiratory epithelium as one aspect of influenza pathogenesis (see also Chapter 5). However, cell lysis does not occur until the cell has produced many thousands of new virus particles.

Cleavage activation of HA and viral pathogenicity

As mentioned above, cleavage of HA0 into HA1 and HA2 is a prerequisite for the expression of HA's membrane fusion activity. Thus, cleavage of HA0 is essential for viral infectivity.[1,11,12,28,29] In human influenza viruses, cleavage is thought to occur extracellularly, at a single arginine residue, after HA0 has been incorporated in virus particles.[37] The enzyme responsible for cleavage, a trypsin-like protease, is probably released from Clara cells in the respiratory epithelium. Since the tissue distribution of this enzyme is limited, human influenza viruses do not normally spread beyond the respiratory tract.

For avian influenza viruses, the situation is somewhat different and appears to be closely related to the pathogenicity of these viruses in birds.[37,38] HAs of non-virulent or low-pathogenicity avian influenza (LPAI) viruses have a single basic cleavage site similar to that in HAs of human viruses, restricting the spread of these viruses. On the other hand, HAs of highly pathogenic avian influenza (HPAI) viruses ("fowl plague" viruses) appear to have a multibasic cleavage site, as a result of which these HAs can be cleaved by intracellular furin-like proteases, in a process that occurs just before the HA molecules reach the surface of the infected cell. Since furin-like proteases are present in virtually every cell type, avian viruses with HAs containing a multibasic cleavage site can easily spread throughout the body, often causing a fatal systemic infection.

It is not clear whether the presence of a multibasic cleavage site in HA0 also represents an important determinant of viral pathogenicity in humans.[37] The HA of the extrememely pathogenic Spanish flu virus (see Chapter 3) does not have the multibasic cleavage site.[39] On the other hand, infection of humans with the highly pathogenic H5N1 avian influenza virus, the HA of which does have the multibasic cleavage site, exhibits a very high case-fatality rate, with six deaths from 18 confirmed cases in Hong Kong in 1997 and 94 deaths among 179 confirmed infections thusfar (March 1, 2006) in Asia and Turkey during the more recent H5N1 outbreaks in these areas. This suggests that the presence of a multibasic cleavage site in HA may also be a determinant of systemic spread and viral pathogenicity in humans, although other factors are probably involved as well.

References

1. Lamb RA, Krug RM. *Orthomyxoviridae*: the viruses and their replication. In: Knipe DM, Howley PM, Griffin DE *et al.*, editors. *Fields Virology*, 4th edn. Lippincott Williams & Wilkins, 2001; pp. 1487–1531.

2. Wright PF, Webster RG. Orthomyxoviruses. In: Knipe DM, Howley PM, Griffin DE *et al.*, editors. *Fields Virology*, 4th edn. Lippincott Williams & Wilkins, 2001; pp. 1533–1579.

3. Moser MR, Bender TR, Margolis HS *et al.* An outbreak of influenza aboard a commercial airliner. *Am J Epidemiol* 1979; **110**: 1–6.

4. Hinshaw VS, Webster RG, Bean WJ, Sriram G. The ecology of influenza viruses in ducks and analysis of influenza viruses with monoclonal antibodies. *Comp Immunol Microbiol Infect Dis* 1980; **3**: 155–164.

5. Osterhaus ADME, Rimmelzwaan GF, Martina BE, Bestebroer TM, Fouchier RAM. Influenza B virus in seals. *Science* 2000; **288**: 1051–1053.

6. WHO Memorandum. A revised system for nomenclature of influenza viruses. *Bull World Health Org* 1980; **58**: 585–591.

7. Ruigrok RWH. Structure of influenza A, B and C viruses. In: Nicholson KG, Webster RG, Hay AJ, editors. *Textbook of Influenza*. Blackwell Science, 1998; pp. 29–42.

8. Steinhauer DA, Wharton SA. Structure and function of the haemagglutinin. In: Nicholson KG, Webster RG, Hay AJ, editors. *Textbook of Influenza*. Blackwell Science, 1998; pp. 54–64.

9. Colman PM. Structure and function of the neuraminidase. In: Nicholson KG, Webster RG, Hay AJ, editors. *Textbook of Influenza*. Blackwell Science, 1998; pp. 65–73.

10. Varghese JN, Laver WG, Colman PM. Structure of the influenza virus glycoprotein antigen neuraminidase at 2.9 Å resolution. *Nature* 1983; **303**: 35–40.

11. Wilson IA, Skehel JJ, Wiley DC. Structure of the haemagglutinin membrane glycoprotein of influenza virus at 3 Å resolution. *Nature* 1981; **289**: 366–373.

12. Skehel JJ, Wiley DC. Receptor binding and membrane fusion in virus entry: the influenza hemagglutinin. *Annu Rev Biochem* 2000; **69**: 531–569.

13. Wiley DC, Wilson IA, Skehel JJ. Structural identification of the antibody-binding sites of Hong Kong influenza haemagglutinin and their involvement in antigenic variation. *Nature* 1981; **289**: 373–378.

14. Hirst GK. Adsorption of influenza haemagglutinins and virus by red blood cells. *J Exp Med* 1942; **76**: 195–209.

15. Palese P, Compans RW. Inhibition of influenza virus replication in tissue culture by 2-deoxy-2,3-dehydro-N-trifluoroacetylneuraminic acid (FANA): mechanism of action. *J Gen Virol* 1976; **33**: 159–163.

16. Von Itzstein M, Wu WY, Kok GB *et al*. Rational design of potent sialidase-based inhibitors of influenza virus replication. *Nature* 1993; **363**: 418–423.

17. Hay AJ. Functional properties of the virus ion channels. In: Nicholson KG, Webster RG, Hay AJ, editors. *Textbook of Influenza*. Blackwell Science, 1998; pp. 74–81.

18. Lamb RA, Zebedee SL, Richardson CD. Influenza virus M2 protein is an integral membrane protein expressed on the infected-cell surface. *Cell* 1985; **40**: 627–633.

19. Zebedee SL, Lamb RA. Influenza A virus M2 protein: monoclonal antibody restriction of virus growth and detection of M2 in virions. *J Virol* 1988; **62**: 2762–2772.

20. Hay AJ. The action of adamantanamines against influenza A viruses: inhibition of the M2 channel protein. *Semin Virol* 1992; **3**: 21–30.

21. Noda T, Sagara H, Yen A *et al.* Architecture of ribonucleoprotein complexes in influenza A virus particles. *Nature* 2006; **439**: 490–492.

22. Weis W, Brown JH, Cusack S *et al.* Structure of the influenza virus haemagglutinin complexed with its receptor, sialic acid. *Nature* 1988; **333**: 426–431.

23. Ito T, Couceiro JN, Kelm S *et al.* Molecular basis for the generation in pigs of influenza A viruses with pandemic potential. *J Virol* 1998; **72**: 7367–7373.

24. Matlin KS, Reggio H, Helenius A, Simons K. Infectious entry pathway of influenza virus in a canine kidney cell line. *J Cell Biol* 1981; **91**: 601–613.

25. Rust MJ, Lakadamyali M, Zhang F *et al.* Assembly of endocytic machinery around individual influenza viruses during viral entry. *Nat Struct Mol Biol* 2004; **11**: 567–573.

26. Smith AE, Helenius A. How viruses enter animal cells. *Science* 2004; **304**: 237–242.

27. Stegmann T, Morselt HW, Scholma J, Wilschut J. Fusion of influenza virus in an intracellular acidic compartment measured by fluorescence dequenching. *Biochim Biophys Acta* 1987; **904**: 165–170.

28. Kim Carr CM, Kim PS. A spring-loaded mechanism for the conformational change of influenza hemagglutinin. *Cell* 1993; **73**: 823–832.

29. Bullough PA, Hughson FM, Skehel JJ, Wiley DC. Structure of influenza haemagglutinin at the pH of membrane fusion. *Nature* 1994; **371**: 37–43.

30. Chen J, Skehel JJ, Wiley DC. N- and C-terminal residues combine in the fusion–pH influenza hemagglutinin HA(2) subunit to form an N cap that terminates the triple-stranded coiled coil. *Proc Natl Acad Sci USA* 1999; **96**: 8967–8972.

31. Helenius A. Unpacking the incoming influenza virus. *Cell* 1992; **69**: 577–578.

32. Bron R, Kendal AP, Klenk HD, Wilschut J. Role of the M2 protein in influenza virus membrane fusion: effects of amantadine and monensin on fusion kinetics. *Virology* 1993; **195**: 808–811.

33. Braakman I, Hoover-Litty H, Wagner KR, Helenius A. Folding of influenza hemagglutinin in the endoplasmic reticulum. *J Cell Biol* 1991; **114**: 401–411.

34. Doms RW, Lamb RA, Rose JK, Helenius A. Folding and assembly of viral membrane proteins. *Virology* 1993; **193**: 545–562.

35. Rodriguez-Boulan E, Sabatini DD. Asymmetric budding of viruses in epithelial monlayers: a model system for study of epithelial polarity. *Proc Natl Acad Sci USA* 1978; **75**: 5071–5075.

36. Fujii Y, Goto H, Watanabe T, Yoshida T, Kawaoka Y. Selective incorporation of influenza virus RNA segments into virions. *Proc Natl Acad Sci USA* 2003; **100**: 2002–2007.

37. Steinhauer DA. Role of hemagglutinin cleavage for the pathogenicity of influenza virus. *Virology* 1999; **258**: 1–20.

38. Klenk H-D, Garten W. Host cell proteases controlling virus pathogenicity. *Trends Microbiol* 1994; **2**: 39–43.

39. Reid AH, Fanning TG, Hultin JV, Taubenberger JK. Origin and evolution of the 1918 “Spanish” influenza virus hemagglutinin gene. *Proc Natl Acad Sci USA* 1999; **96**: 1651–1656.

3. Influenza Epidemics and Pandemics

Influenza viruses continuously undergo antigenic evolution.[1] It is this quality that allows them to evade the pre-existing immunity of the host, which implies that immune responses mounted against earlier forms of the virus are less effective or completely ineffective against newer variants. Since the viral surface glycoprotein HA is the antigen against which virus-neutralizing antibodies are directed (see Chapter 4), it is primarily the antigenic variation of HA that is responsible for the immune escape of influenza viruses. Other viral antigens undergo significantly less variation, but – under immunological pressure – may also contribute to the evasion of the immune defence of the host.[2] The antigenic evolution of influenza viruses forms the primary basis for the occurrence of annual influenza epidemics and occasional pandemics. Also, it necessitates regular updates of the composition of the influenza vaccine (see Chapter 8).

There are two main mechanisms by which influenza A viruses change their antigenic properties. These are commonly referred to as "antigenic drift" and "antigenic shift".[1] Antigenic drift occurs through continuous mutation of the RNA genome of the virus. As a result, amino acid changes in the viral proteins arise, including substitutions in the antigenic epitopes of HA. Virus variants with such substitutions have a selective advantage over the original virus, since they are no longer or less efficiently countered by the immunity of the host.

Occasionally, an entirely new influenza A virus subtype of avian origin emerges in the human population. This is called antigenic shift. Since there is no immunity whatsoever to the new virus subtype in the population, this shift forms the scenario for a pandemic outbreak of influenza. Antigenic shift may be the result of direct transmission of an avian virus to humans. Alternatively, a human influenza virus may acquire a subset of genes from an avian virus, pigs or humans serving as a "mixing vessel". This process of genetic "reassortment" (see Chapter 2) might result in the formation of a new virus with most of the properties of a normal human influenza virus, yet having a new HA on its surface to which the population is immunologically naive.[1]

Three major influenza pandemics struck the world in the 20th century. By far the most devastating pandemic was the Spanish flu outbreak, which hit at the end of the First World War.[3,4] It spread across the globe in three consecutive waves in 1918–19, killing at least 50 million people.[5] The subsequent pandemics in 1957 and 1968 were milder,[4] but nonetheless also caused a total of approximately 2 million deaths. There is no doubt that there will be other pandemics in the future.[6,7] Having its reservoir among migratory waterfowl, the influenza virus is non-eradicable and will continue to affect humans. The recent outbreaks of highly pathogenic avian influenza, and the occasional transmissions to humans with a fatality rate of over 50%, should serve as a warning that the emergence of a new pandemic is just a matter of time. Accordingly, the WHO is urging individual countries to be aware of the need to start formal influenza pandemic preparedness planning as soon as possible.

Annual influenza epidemics

Nature of antigenic drift

The replication of viral RNA in an infected host cell is an error-prone process, much more so than the replication of DNA. One reason for this is the lack of so-called proofreading mechanisms in the replication of RNA. As a result of the infidelity of RNA replication, the mutation frequency in the influenza virus RNA is approximately one in 100,000 nucleotides.[8] Considering that the size of the entire influenza virus genome is about 14,000 nucleotides, this implies that many new viral RNA genome copies will contain one or more mutations. While the majority of these

Key Messages
• Influenza viruses continuously change their antigenic properties through "antigenic drift" and "antigenic shift".
• Antigenic drift is due to accumulation of mutations in antigenic epitopes of viral antigens. As a result, pre-existing antibodies against the viral HA neutralize newer drift variants of the virus less efficiently.
• Antigenic drift of HA is the cause of recurrent annual influenza epidemics. It necessitates repeated vaccination of target groups and regular updates of the vaccine composition.
• The natural reservoir of influenza A viruses is among wild aquatic birds. All 16 subtypes of HA and all nine subtypes of NA have been identified in waterfowl.
• Antigenic shift implies that an influenza A virus with a new HA subtype is introduced into the human population. This may occur by (i) direct transmission of an avian virus from birds to humans, (ii) genetic reassortment between an avian and a human virus, or (iii) reintroduction of an "old" strain into the population.
• Antigenic shift is the cause of occasional global outbreaks of influenza (pandemics). These arise primarily because of a total lack of immunity against the new virus subtype in the population.
• Since the reservoir of influenza A viruses is among wild birds and avian viruses do occasionally cross the species barrier to humans, influenza is a non-eradicable disease and pandemics will continue to occur.
• Intensive surveillance of influenza activity among birds (and other species, including humans) is required for optimal pandemic preparedness.
• To improve the global level of pandemic preparedness, the WHO is encouraging individual countries to institute pandemic preparedness plans and to stimulate annual influenza vaccine uptake.

mutations will be silent or generate stop codons, others will result in amino acid substitutions in the translated proteins, which may give rise to new variants with an advantage over the parent virus. Antigenic drift of HA occurs as a result of accumulation of point mutations in the antigenic domains (epitopes) of HA1 (see Figure 7). Selective advantage for the new variants in this case arises because these variants are no longer neutralized by pre-existing antibodies of the host.

Antigenic drift is a gradual process. Mutations may arise in one epitope of HA with other epitopes remaining unchanged. Thus, depending on the nature of the mutation, the new strain may still be partially recognized by the immune defence of the host. This gradual change is illustrated in Table 3, which shows that drift variants of the 1968 Hong Kong pandemic virus gradually lost the ability to cross-react with an antiserum against the original virus, while antisera against increasingly distant drift viruses no longer cross-react with the 1968 strain.[1]

Seasonality of influenza outbreaks

While in tropical countries influenza may be present all year round, epidemics in temperate regions occur almost exclusively in the winter months, from October to April in the northern hemisphere and from April to October in the southern hemisphere. It is not clear why there is this seasonal regularity in the occurrence of influenza outbreaks. One reason is thought to be crowding of people in closed spaces with poor ventilation. For example, spreading of the virus may be facilitated in schools – there is evidence that school children in particular play a prominent role in the transmission of influenza.[9]

Table 3. Antigenic drift of influenza H3N2 virus from 1968 to 1997. The table provides haemagglutination–inhibition titres (see Chapter 8) of ferret antisera against drift variants of the H3N2 virus. Reproduced from Wright PF, Webster RG. Orthomyxoviruses. In: Knipe DM, Howley PM, Griffin DE *et al.*, editors. *Fields Virology*, 4th edn. Lippincott Williams & Wilkins, 2001; pp. 1533–1579[1] with permission from Lippincott Williams & Wilkins.

Antigenic drift of influenza H3N2 virus from 1968 to 1997										
	Reference ferret antisera									
Strain designation	HK/68 1.	ENG/72 2.	VIC/75 3.	TEX/77 4.	BANG/79 5.	PHIL/82 6.	MISS/85 7.	SHN/87 8.	BEI/89 9.	BEI/92 10.
1. A/Hong Kong/01/68	*2560*	1280	40	5	5	5	10	10	5	5
2. A/England/42/72	320	*1280*	80	20	5	5	5	5	5	5
3. A/Victoria/03/75	5	80	*320*	40	10	20	10	5	5	5
4. A/Texas/01/77	5	80	160	*1280*	160	320	320	10	10	5
5. A/Bangkok/01/79	5	5	80	640	*640*	640	640	20	10	5
6. A/Philippines/02/82	5	5	10	40	40	*320*	160	10	10	5
7. A/Mississippi/01/85	5	5	40	160	80	640	*640*	40	20	20
8. A/Shanghai/11/87	5	5	5	5	5	5	80	*320*	160	20
9. A/Beijing/353/89	5	5	5	5	5	5	5	160	*320*	40
10. A/Beijing/32/92	5	5	5	5	5	5	10	20	80	*640*
11. A/Johannesburg/33/94	5	5	5	5	5	5	5	10	20	160
12. A/Nanchang/933/95	5	5	5	5	5	5	5	5	10	20
13. A/Sydney05/97	5	5	5	5	5	5	5	5	5	5

Despite their annual seasonal character, influenza epidemics are unpredictable.[4,7] When precisely they will start and how long they will last are questions that are difficult to answer in advance. Figure 11 illustrates the variation in the onset and duration of influenza epidemics recorded in the Netherlands in the last three decades of the previous century. It is also impossible to predict how virulent a winter epidemic will be. The severity of an epidemic in any given year is a subtle interplay between the waning immunity in the population, the extent of antigenic drift of the virus and the intrinsic virulence of the new virus variant.

Regular update of vaccine composition and influenza surveillance

The antigenic drift of influenza virus necessitates regular updates of the composition of the influenza vaccine (see Chapter 8). Obviously, a close match of the virus strains in the vaccine and the strains subsequently circulating in the winter season is of key importance for the protective efficacy of the vaccine. Current inactivated influenza vaccines contain three virus strains (two A strains and one B strain), the specific composition of the vaccine being recommended annually by the WHO.

In 1947, the WHO established an international Influenza Surveillance Network of laboratories to monitor the emergence and spread of new influenza strains around the world and thus to predict the circulating strains to be represented in the vaccine. This activity is co-ordinated by the WHO Collaborating Centres based in Atlanta, London, Melbourne and Tokyo. In general, despite the inherent uncertainty of this approach, the selection of vaccine virus strains has worked quite well thus far (see Chapter 8). In addition, the uncertainty in predicting epidemic virus strains has been further reduced by mathematical modeling of the antigenic distance between different viruses, allowing a more accurate selection of the strains that need to be included in the vaccine.[10] However, when occasional mismatches do occur, people in target groups should

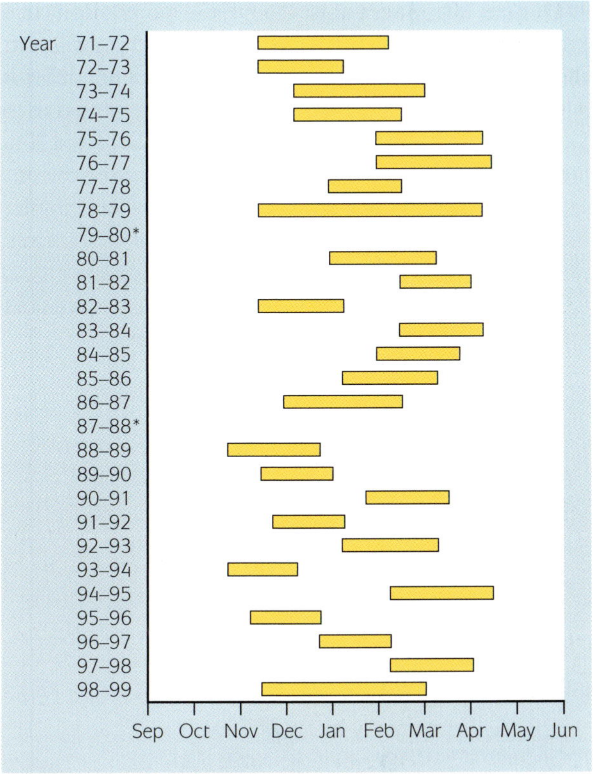

Figure 11. Onset and duration of influenza epidemics in the Netherlands, 1971–1999. Courtesy of Solvay Pharmaceuticals, Weesp, the Netherlands.

nonetheless be encouraged to be vaccinated, as there is still benefit in spite of the reduced efficacy of the vaccine under these conditions.

Origin of influenza pandemics

Antigenic shift

Currently available information suggests that the natural reservoir of influenza A viruses is among aquatic birds.[11–13] All 16 HA and nine NA subtypes have been identified in birds (see Chapter 2). Occasionally, a new influenza virus subtype is introduced into the human population, in a process referred to as antigenic shift.[14] As indicated above, there are several ways by which such new human virus subtypes may arise (Figure 12), and it is likely that each has played a role in the influenza pandemics of the 20th century. First, an avian virus may be transmitted directly

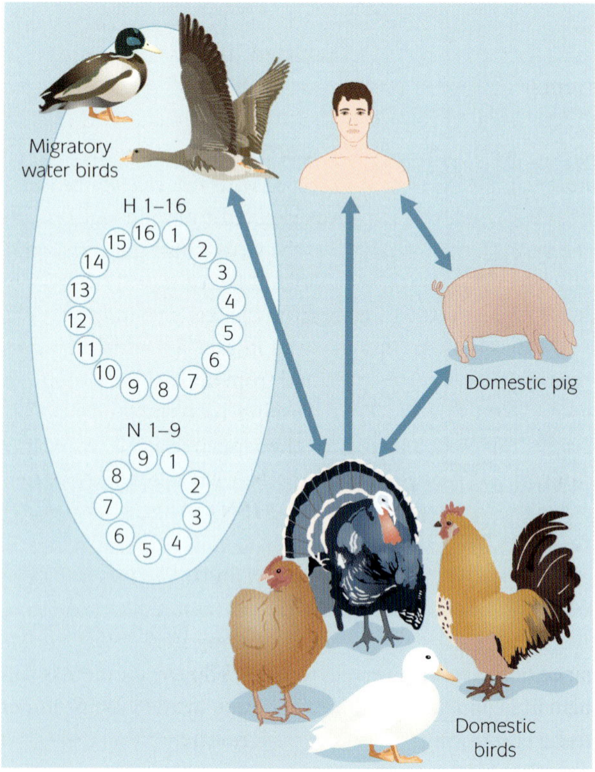

Figure 12. Origin of antigenic shift and pandemic influenza. The natural reservoir of influenza A viruses is among migratory aquatic birds. A virus with a new HA subtype may be introduced into the human population by direct transmission of an avian virus to humans or by genetic reassortment between an avian and a human virus. Adapted from Nicholson KG *et al*. Influenza. *Lancet* 2003; **362**: 1733–1745[7] with permission from Elsevier.

to humans and subsequently adapt to the new host by mutation. Second, a human influenza virus may acquire a number of gene segments from an avian virus in a process of genetic reassortment. Third, it is possible that an "old" human virus, which has circulated before, is reintroduced into the human population.

Direct transmission of avian viruses to humans

Rarely, a highly pathogenic avian influenza (HPAI; see also Chapter 2) virus is transmitted directly from birds to humans.[3,11,15–18] It was not until the 1997 H5N1 bird flu outbreak in Hong Kong that it was appreciated that such direct transmission could occur. During this outbreak, 18 people were reportedly infected with the H5N1 virus, six of whom died.[18–20] The virus isolated from the victims had the same genetic make-up as the virus causing the epidemic among chickens in Hong Kong at the time, indicating that it was transmitted directly from chickens to humans. The 1997 H5N1 bird flu virus was in fact a "multiple avian reassortant", which originated from an H5N1 reservoir in geese, reassorted before crossing to terrestrial poultry (especially chickens), and then probably acquired the internal protein genes and the NA gene through further reassortment with H9N2 and H6N1 viruses originating from quails.[21]

While the bird flu outbreak in 1997 was the first documented example of a purely avian virus causing respiratory disease and deaths among humans,[18–20] it has since become apparent that the 1918 Spanish flu virus was also an avian virus.[3,22–25] Even though the precise origin of the 1918 virus remains enigmatic,[3,11,22] based on characterization of genetic material isolated from victims of the 1918 pandemic, it is clear now that the virus was not a human–avian reassortant, but rather an avian-like virus that was introduced in its entirety into the human population and subsequently adapted to the new host. The adaptation involved a combination of mutations in different genes of the virus, including the HA and polymerase genes.[3,23–25] Adaptation of an avian virus to the human host might thus give rise to a virus capable of human-to-human transmission. Determination of the structure and receptor-binding properties of the 1918 HA[24] has suggested that, indeed, by mutation of the receptor-binding site, the HA acquired the ability to interact with human α2,6-linked sialic acid receptors (see Chapter 2). This may explain how the virus evolved the ability to spread efficiently in the human population, although other viral genes were likely involved in human-to-human transmission as well. Since adaptation to the new host requires the accumulation of a number of mutations, it may involve a considerable period of time. Indeed, there is some circumstantial evidence that the 1918 virus entered the human population several years before the pandemic.[3,22] There is a close genetic relationship between the 1918 virus and the porcine A/Swine/Iowa/30 (H1N1) virus. However, this does not necessarily imply that the human virus originated in pigs. Rather, the 1918 virus or its precursor may have been transmitted from humans to pigs.[4,22]

Remarkably, the H5N1 virus isolated from victims of the recent outbreaks in Asia have acquired a number of mutations in the polymerase complex that are also present in the 1918 virus.[25] These mutations are likely important for efficient replication of the virus in humans. Indeed, once having crossed the species barrier to humans, the H5N1 evidently replicates very efficiently and is highly pathogenic (see Chapter 5), killing about half of the people with a confirmed H5N1 infection. Fortunately, there is no evidence yet for human-to-human transmission of the virus, all reported cases of human infection being directly linked to exposure of the victims to infected poultry. Clearly, additional mutations or reassortment events are required to permit efficient human-to-human spread. If, however, the H5N1 virus would acquire the ability to efficiently spread in the human population, while maintaining its current level of pathogenicity, the scale of the ensuing global catastrophe would be beyond comprehension.[11]

Genetic reassortment

Genetic reassortment may occur when a host cell is simultaneously infected with two influenza A viruses. The RNA segments from both viruses are replicated in the nucleus of the cell. During the reassembly of new virus particles (see Chapter 2), the RNA segments from the two strains can get mixed together and a third "new" viral strain can be produced with a unique combination of genes.

As illustrated in Figure 13, the 1957 Asian H2N2 virus subtype obtained its HA, NA, and PB1 genes from an avian virus and the other fives genes from the circulating H1N1 human strain,[1,4,7,14–17] the human H1N1 virus in turn being a distant descendent of the 1918 Spanish flu virus.[3] Likewise, the 1968 Hong Kong H3N2 virus acquired its HA and PB1 genes from an avian virus, but retained the NA and remaining five other genes from the then circulating H2N2 human virus. It is remarkable that the H3 of the Hong Kong virus differed in only seven amino acid residues from its hypothetical avian ancestor virus, providing very strong support for the concept that the Hong Kong virus indeed arose as a result of genetic reassortment.[26]

It has long been thought that genetic reassortment would occur exclusively in pigs. The pig provides an ideal "mixing vessel", since pigs are readily infectable by both human and avian influenza viruses due to the molecular nature of their sialic acid receptors[27] (see Chapter 2). However, there is increasing evidence now that other species, including man, might also serve as mixing vessels. Indeed, since HPAI viruses do have the ability to occasionally cross the species barrier to humans, coinfection of cells with an avian and a human virus may well occur within a human host.

Reintroduction of an old strain

A third way by which an antigenic shift may occur and a pandemic arise is reintroduction of an "old" strain into the human population. A virus could remain hidden, for example in a frozen state such that little change would occur, and re-

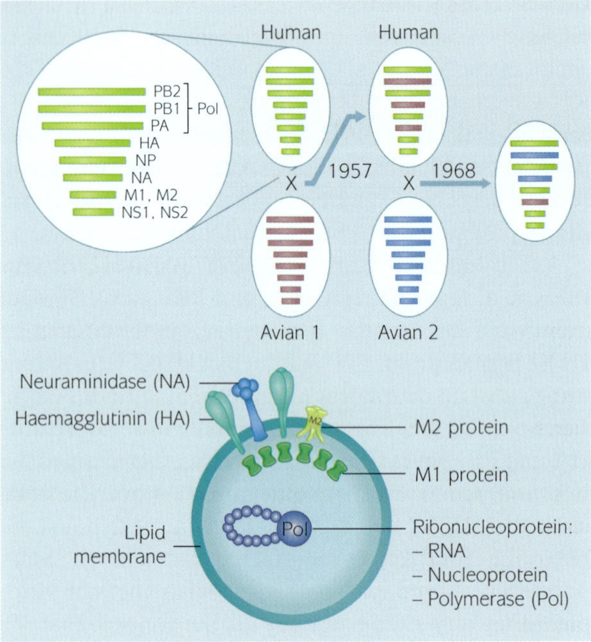

Figure 13. Genetic reassortment between human and avian influenza A viruses as the origin of the pandemic 1957 Asian flu and 1968 Hong Kong flu viruses. The segmented nature of the influenza virus genome facilitates reassortment. Pigs – which support the replication of both avian and human influenza A viruses – possibly serve as a "mixing vessel". The Asian H2N2 virus subtype obtained its HA, NA and PB1 genes from an avian virus. Likewise, the HA and PB1 genes of the Hong Kong H3N2 virus are of avian origin – the Hong Kong virus retained the NA gene plus five other genes from the circulating human virus. Adapted from Potter CW, editor. *Influenza (Perspectives in Medical Virology)*, 2003 with permission from Elsevier.

emerge as a pandemic virus when the immunity in the population to that virus has waned. The appearance of the Russian flu (H1N1) in 1977 supports the idea of reintroduction of a previous strain. The virus turned out to be identical in all of its genes to the H1N1 virus that circulated before 1957 and caused a major epidemic in 1950.[4,28]

Influenza pandemics of the past

Early records of influenza outbreaks

Historic records indicate that influenza pandemics have been among us for many centuries.[29] A report by Hippocrates describes an outbreak as early as 412 BC that could have been due to influenza. Several further accounts are suggestive of influenza outbreaks, but lack sufficient scientific substance to provide certainty. The first convincing report of an influenza pandemic describes an outbreak in 1580, which started in Asia in the summer, and then spread via Africa to Europe and finally to America.[26] In Britain it came in two waves, a pattern that was also seen in subsequent influenza pandemics. Other convincing accounts of influenza pandemics followed in thc 18th and 19th centuries.[29]

Pandemics of the 20th century

Figure 14 gives an overview of the influenza pandemics that occurred in the past century.[4,29,30] The most terrible outbreak was the Spanish flu in 1918, with an estimated 50 million deaths,[5] justifying its description as "the last great plague of mankind".

1918 – Spanish flu (H1N1)

The Spanish flu first struck in March 1918 in the USA, with reports from Detroit, South Carolina and San Quentin Prison of outbreaks of an unusual respiratory disease that was associated with a disproportionate increase in deaths among young adults. The infection spread rapidly, crossing both the Atlantic and Pacific Oceans in a matter of months, killing many American soldiers during their voyage to Europe in the spring of 1918 (Figure 15). In Europe, initial infections were reported in Madrid in May 1918, hence the name "Spanish" flu. This first wave infected all the armies in Europe during May and June. After the initial wave there was a sharp decline

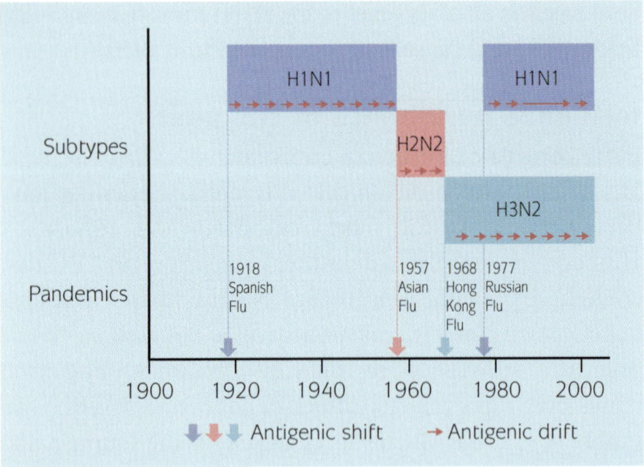

Figure 14. Influenza A viruses of the 20th century pandemics. The 1918 Spanish flu virus (H1N1) probably originated through direct transmission of an avian virus to man and further adaptation of this virus to the human host. The 1957 Asian (H2N2) and 1968 Hong Kong (H3N2) viruses were reassortants of avian and circulating human viruses (see Figure 13). The 1977 Russian flu virus was an old strain, reintroduced into the human population. Courtesy of Solvay Pharmaceuticals, Weesp, the Netherlands.

of infections in the summer of 1918, to be followed by a second, more severe, wave that peaked in the autumn of 1918, and a third wave in the spring of 1919, both associated with global spread of the infection. In just 10 months, approximately 50 million people worldwide were killed,[5] more than the total number of victims of the 1914–18 war.

The origin of the Spanish influenza is not clear. It may have been imported into the USA by migrant workers from China. It is equally possible that the pandemic originated in the USA. Indeed, the fact that the pandemic started with simultaneous outbreaks at three different locations in the USA supports this hypothesis.

Recently, based on fragments of genetic material isolated from the remains of a victim of the Spanish flu pandemic buried in the permafrost of Alaska, the 1918 virus has been

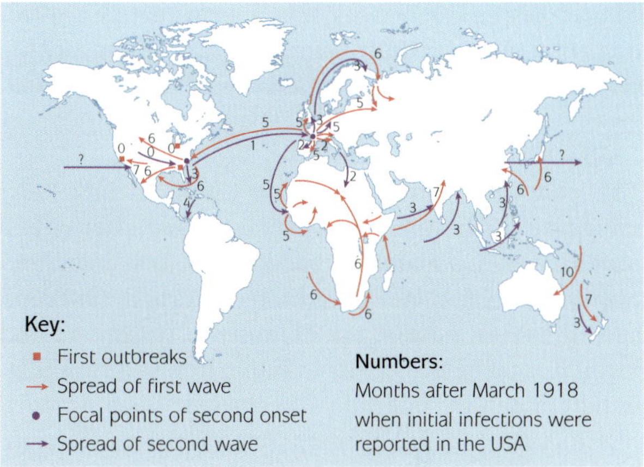

Figure 15. The 1918 Spanish flu pandemic. The origin of the Spanish influenza virus (H1N1) is unclear. First outbreaks were reported in the USA in March 1918. The infection rapidly spread to Europe, where its impact was particularly terrible, and further throughout the world. After the first wave in the spring of 1918, a second, more devastating wave followed in the fall of 1918 and a third wave in the spring of 1919. The pandemic killed an estimated 50 million people. Adapted from Potter CW. Chronicle of influenza pandemics. In: Nicholson KG, Webster RG, Hay AJ, editors. *Textbook of Influenza*. Oxford: Blackwell Science, 1998; pp. 3–18[29] with permission from Blackwell Publishing.

resurrected,[31] using reverse genetics techniques (see Chapter 9). Challenge studies in mice have confirmed the extreme lethality of the virus.

1957 – Asian flu (H2N2)

The "Asian flu" originated in the Yunan Province of China in March 1957, spreading rapidly to South-East Asia and Japan, and subsequently in May to Australia, Indonesia and India, and during the summer to Europe, Africa, North and South America, and the Caribbean. In just 6 months, the pandemic had spanned the globe. A second wave occurred during the autumn of 1957. Altogether it affected some 40–50% of the population, with 25–30% experiencing clinical disease. Mortality was approximately 1/4000, occurring predominantly among the very young and the very old. The total death rate probably exceeded 1 million.

1968 – Hong Kong flu (H3N2)

This pandemic also originated in China, in July 1968, spreading to Hong Kong, where it peaked after only 2 weeks, causing a major outbreak of 500,000 cases, and receded completely after 6 weeks. By August, the infection had spread to Taiwan, the Philippines, Singapore and Vietnam, and by September to India, Iran and Australia. In that month the infection also entered North America via California, carried by troops returning from Vietnam. In the USA, the epidemic peaked in December. Altogether, 30–40% of the population were infected, causing 56,000 excess deaths. School absenteeism reached 50%. Compared to the earlier pandemics of the century, the Hong Kong flu was relatively mild, possibly as a result of immunity in the population against the N2 subtype neuraminidase which the new pandemic H3N2 virus shared with the circulation H2N2 virus. Worldwide, the total death toll of the 1968 pandemic was between 500,000 and 1 million.

1977 – Russian flu (H1N1)

The outbreak of Russian flu first appeared in northen China in May 1977 and spread throughout Russia by December, and the rest of the world in 1978. As mentioned above, the virus was subsequently found to be virtually identical to one that had caused a human epidemic in 1950.[28] Consequently, most people over 23 years old possessed antibody to it. Thus, the pandemic was confined almost entirely to children and teenagers. Thankfully, the illness was quite mild and weekly attack rates at peak were about 13% in children 7–14 years old. Unlike the previous two pandemic viruses, this virus failed to replace the previously circulating influenza A virus, such that currently both H1N1 and H3N2 viruses circulate in humans.

Will there be a new pandemic?

There is no doubt that there will be influenza pandemics in the future.[7,11,22] It is not so much a matter of *whether* they will occur but rather *when* they will occur. All the ingredients for the formation of new influenza viruses with pandemic potential are there, particularly in Asia. As indicated above, the 1957 H2N2 and 1968 H3N2 viruses originated in China. It is thought that the ecological conditions in China, as well as many other Asian countries, are such that new viruses with pandemic potential may arise comparatively easily. These include the year-round circulation of influenza viruses, along with the dense populations of people, pigs and domestic poultry, often living in close proximity – conditions that facilitate genetic reassortment or direct transfer of avian viruses to humans. However, new viruses with pandemic potential could also arise elsewhere. In fact, human–avian reassortant viruses have been found in Italy and the Netherlands. Also, the 2003 H7N7 fowl plague outbreak in the Netherlands, with a fatal human infection (see below), and the ongoing outbreaks of H5N1 not only in Asia but also in Europe, Africa and the Middle East, underline the risk of reassortment or adaptation to humans, and indicates that new viruses with pandemic potential need not necessarily arise in Asia. Also, previous pandemic human virus subtypes (e.g. the H2N2 virus) may re-emerge when the immunity in the population against such viruses has gradually disappeared. It is likely that this has happened frequently in the past.

It is not possible to predict when the next pandemic will strike. Since 1889, pandemics have occurred at intervals ranging from 10 to 40 years. A cyclic theory has been proposed suggesting that pandemics may appear in a specific recurring pattern.[28] The interval between the 1889 (H2) and 1900 pandemics (H3) matches that between the 1957 (H2) and 1968 (H3) pandemics. However, the approximate 20-year interval between the 1900 and 1918 pandemics would have predicted a global outbreak around 1988, 20 years after the 1968 pandemic, but this did not occur.

"Near misses" of the recent past

"H5N1" (1997–2006)

It is widely assumed that the mass slaughter of 1.5 million chickens in December 1997 in the Hong Kong H5N1 outbreak may well have averted a new human pandemic. Since this incident, the H5N1 virus has re-emerged several times. There is evidence that the precursor viruses of the H5N1 bird flu virus, including the goose H5N1 virus, remained in circulation.[32] Since 1998, culling has also included geese to prevent the goose H5N1 precursor virus from taking a firm hold in poultry in Hong Kong.

In May 2001, and in February and April 2002, the H5N1 virus subtype was once again detected in Hong Kong's poultry markets,[7,33] the specific virus strain differing significantly from the 1997 H5N1 bird flu virus. It had apparently crossed from geese to ducks, undergone reassortment and then crossed to other poultry, but at the time it caused no human infections. However, in February 2003, a 9-year-old boy was hospitalized with an influenza A H5N1 illness, which also affected other members of his family. He survived but his father and 8-year-old sister died. Then, in the spring of 2004, there was a major outbreak of highly pathogenic H5N1 avian influenza first detected in southern Vietnam and then spreading quickly to neighbouring countries, and quite recently also to Europe and Africa.[34–36] To date (March 1, 2006) there have been 179 confirmed cases of human H5N1 infections, with 94 fatalities, including four deaths in Turkey in January 2006.[37] This high case-fatality rate underscores the extremely high pathogenicity of the H5N1 virus involved.

"H9N2" (1999)

In southern China, nine cases of human infection with an H9N2 avian virus were reported in 1998. In March 1999, there were two further cases in Hong Kong.[38] In no case was there serological evidence for H9N2 infection among family members or health-care workers in close contact with the infected individuals, indicating that there was no human-to-human transmission. The virus was similar to one of the putative precursor viruses of the 1997 outbreak in Hong Kong. It appears that the H9N2 viruses are also circulating in pigs, together with human and porcine H1N1 and H3N2 viruses, providing conditions required for the generation of a reassortant virus with pandemic potential in humans.

"H7N7" (2003)

In the Netherlands in 2003, an outbreak of H7N7 fowl plague affected mainly poultry workers and veterinarians, causing conjunctivitis in most cases and influenza-like illness in about 10%.[39] There was one death – a 57-year-old veterinarian who visited one of the infected farms and later died of acute respiratory distress syndrome. Mass culling of affected poultry was instigated and agricultural workers in contact with affected poultry were treated prophylactically with antiviral drugs. Also, these poultry workers were vaccinated for the human influenza strains circulating at the time to minimize dual infections as an opportunity for genetic reassortment.

Influenza pandemic preparedness

Since the 1997 H5N1 outbreak in Hong Kong there has been a growing recognition that global and national plans should be made and implemented to minimize the potential implications of the next pandemic. In 1999, the WHO issued its first pandemic preparedness plan,[40] which was updated in 2005.[41] In 2002, the WHO adopted the Global Agenda on Influenza Surveillance and Control as a strategic approach to increase the level of global pandemic preparedness.[42,43] This global agenda sets out a series of activities around four primary objectives: (i) strengthening surveillance; (ii) improving knowledge of disease burden; (iii) increasing vaccine usage; and (iv) accelerating pandemic preparedness.

Surveillance and phasing of the pandemic

A critical requirement for pandemic control is the ability to detect viruses of pandemic potential at an initial stage of their emergence. Therefore, surveillance programmes for early detection of virus activity and infection in animal reservoirs and human populations are in place and are currently being expanded. It is anticipated that the activities under the Global Agenda will facilitate a rapid detection of unusual influenza activity. Hopefully, early detection along with a swift institution of appropriate measures may allow us to avert a pandemic before it starts, or at least lessen its impact.

Another important aspect of pandemic preparedness relates to an unambiguous phasing of a pandemic when it occurs. Clearly, a number of conditions must be satisfied in order for an outbreak of influenza to be classified as a "pandemic".[14] These include: (i) the emergence of a new influenza A virus subtype in the human population, unrelated to the virus subtypes that circulated immediately before the outbreak; (ii) a total lack of immunity to the new virus among the population; (iii) capacity of the virus to spread by person-to-person transmission; and (iv) rapid dissemination of the infection beyond its site of origin. Based on these definitions, the WHO has developed an algorithm to identify specific phases in the period preceding a pandemic and during the actual pandemic, which require different levels of alertness and actions.[41] These phases are indicated in Table 4.

Improvement of knowledge about burden of disease

In many countries, influenza is not a prominent item on the national health agenda due to other health priorities. As a consequence, in these countries, there is no or an insufficient infrastructure for influenza virus and disease surveillance, and a lack of knowledge about the burden of disease. Without a proper infrastructure for epidemic influenza control, the opportunities for effective pandemic containment will be much more limited than in the presence of such infrastructure. Therefore, the WHO currently intensifies and expands its viral surveillance programme to these countries and promotes studies to establish the burden of disease. It is anticipated that the outcome of such studies might be a natural incentive to improve the infrastructure

Specific pandemic phases defined in WHO global influenza preparedness plan
Interpandemic period • *Phase 1:* No new influenza virus subtypes have been detected in humans. An influenza virus subtype that has caused human infection may be present in animals. If present in animals, risk of human infection or disease is considered to be low. • *Phase 2:* No new influenza virus subtypes have been detected in humans. However, a circulating animal influenza virus subtype poses a substantial risk of human disease.
Pandemic alert period • *Phase 3:* Human infection(s) with a new subtype, but no human-to-human spread, or at most rare instances of spread to a close contact. • *Phase 4:* Small cluster(s) with limited human-to-human transmission but spread is highly localized, suggesting that the virus is not well adapted to humans. • *Phase 5:* Larger cluster(s) but human-to-human spread still localized, suggesting that the virus is becoming increasingly better adapted to humans, but may not yet be fully transmissible (substantial pandemic risk).
Pandemic period • *Phase 6:* Pandemic: increased and sustained transmission in general population.

Table 4. Definition of successive pre-pandemic and pandemic phases. Source: WHO global influenza pandemic preparedness plan. The role of WHO and recommendations for national measures before and during pandemics.[41]

for the control of epidemic influenza and increase the level of pandemic preparedness in the countries concerned. During recent international meetings on global pandemic preparedness, strategies have been developed to effectively implement an action plan to achieve the necessary increased level of epidemic and pandemic influenza control.[44,45]

Stimulation of vaccine uptake and development of pandemic vaccines

According to the WHO, one important way to achieve a considerably improved level of pandemic preparedness is to encourage the annual uptake of influenza vaccines.[42,46] Following the logic of "demand drives supply" for seasonal influenza vaccines, an increased annual use of influenza vaccines, according to existing recommendations, will be followed by investments from the vaccine industry to increase their vaccine production capacity. Clearly, an increased global vaccine production capacity will lead to a much improved worldwide vaccine availability in the case of a pandemic. At the same time increased annual vaccine use will reduce the yearly burden of influenza (see also Chapters 6 and 8).

There is also an urgent need for development of more effective vaccine formulations, such that the dose of antigen per vaccination may be reduced and a more equitable global distribution of scarce pandemic vaccine may be achieved (see also Chapter 9). Recent studies with candidate H5N1 vaccines have demonstrated the need for adjuvants to be included in pandemic vaccine formulations. Non-adjuvanted H5N1 vaccines need a very high dosing to elicit adequately protective antibody titres in unprimed individuals. This would dramatically reduce the volume of effective vaccine doses which can be produced and distributed given the current global vaccine production capabilities. At a technical meeting organised by the WHO in 2005, all major influenza vaccine manufacturers have indicated that they are developing adjuvanted pandemic vaccine candidates.[47] In addition, early registration of pandemic influenza vaccine formulations will considerably expedite the availability of vaccine during a pandemic.[48]

Improvement of pandemic preparedness

Pandemic preparedness should be achieved at global, regional and national levels.[41] The WHO and other global bodies such as the FAO (UN Food and Agriculture Organization) and OIE (World Organization for Animal

Health) are responsible for the infrastructure at a global level, whereas national health authorities are primarily responsible for the infrastructural measures within individual countries. The WHO has provided recommendations for appropriate measures in the regions and the countries to increase pandemic preparedness. Apart from the general infrastructural measures to cope with national emergency situations, the level of pandemic preparedness in countries will depend very much on their ability to have timely access to antiviral agents and pandemic vaccines.

The WHO has made recommendations for stockpiling antivirals and H5N1 vaccines for immediate use at the onset of a potential pandemic outbreak.[41,46] It has also advised national health authorities to make pre-pandemic purchase agreements with vaccine manufacturers for pandemic vaccines.[46] It will take at least 4–6 months after the WHO has announced the pandemic before the first pandemic vaccine doses will become available for distribution. Countries with pre-pandemic purchase agreements will be the first beneficiaries of the scarce volumes of pandemic vaccine. Although there is a general moral feeling that there should be an equitable distribution of scarce antivirals and pandemic vaccines, unfortunately, there is currently no proper mechanism in place in the world to achieve such a fair distribution.

One way to achieve a more equitable distribution of pandemic vaccine is to expand the global vaccine production capacity. Indeed, industry, sometimes with the help of national governments, is expanding its production capacity, while several countries are currently considering to build influenza vaccine production facilities on their soil. Although expansion of global vaccine production capacity is needed and must be stimulated, it should also be noted that once these facilities are operational, the risk of global interpandemic overproduction capacity may become realistic. Therefore, strategies to address the concerns about pandemic vaccine shortages should also take into account the need for a sustainable interpandemic influenza vaccine market.

As an integral part of pandemic preparedness planning, a "pandemic simulation" project "Common Ground" was carried out in the European Union in November 2005.[49] The objective was to test the level of preparedness of each of the member states and identify gaps in their respective national pandemic preparedness plans. During this exercise, all stakeholders, including not only representatives of national health authorities and national influenza centres, but also the WHO, the European Medicines Agency (EMEA) and the European Centre for Disease Control (ECDC), as well as the vaccine and antiviral industry, had to deal with three different stages of a simulated pandemic. This highly educational exercise, which is being formally and fully evaluated, revealed the need to improve communication lines between the different stakeholders. Similar pandemic simulations are planned elsewhere in the near future.

Concluding remarks

The realistic threat of a new influenza pandemic has triggered a series of national and international efforts. The increased efforts on surveillance, pandemic vaccine research, production expansion of antivirals and vaccines, and pandemic preparedness planning at national and international levels will all contribute to contain the disastrous effects of such an emergency as much as possible. These activities are carried out by many stakeholders such as the WHO, national health authorities, non-governmental organizations and the vaccine industry. Although the world is not yet fully prepared for the next pandemic and still much has to be done, current intense global efforts justify some prudent optimism about our increased capabilities for early detection and warning of the next pandemic as well as active interventions so that, when the pandemic strikes, the world will be as reasonably prepared as possible and ready to contain the global damage as much as possible.

References

1. Wright PF, Webster RG. Orthomyxoviruses. In: Knipe DM, Howley PM, Griffin DE *et al.*, editors. *Fields Virology*, 4th edn. Lippincott Williams & Wilkins, 2001; pp. 1533–1579.

2. Thomas PG, Keating R, Hulse-Post DJ, Doherty PC. Cell-mediated protection in influenza infection. *Emerg Infect Dis* 2006; **12**: 48–54.

3. Taubenberger JK, Morens DM. 1918 influenza: the mother of all pandemics. *Emerg Inf Dis* 2006; **12**: 15–22.

4. Kilbourne ED. Influenza pandemics of the 20th century. *Emerg Inf Dis* 2006; **12**: 9–14.

5. Johnson NP, Mueller J. Updating the accounts: global mortality of the 1918–1920 "Spanish" influenza pandemic. *Bull Hist Med* 2002; **76**: 105–115.

6. Webby RJ, Webster RG. Are we ready for pandemic influenza? *Science* 2003; **302**: 1519–1522.

7. Nicholson KG, Wood JM, Zambon M. Influenza. *Lancet* 2003; **362**: 1733–1745.

8. Stech J, Xiong X, Scholtissek V, Webster RG. Independence of evolutionary and mutational rates after transmission of avian influenza viruses to swine. *J Virol* 1999; **73**: 1878–1884.

9. Fox JP, Cooney MK, Hall CE, Foy HM. Influenza virus infections in Seattle families, 1975–1979. II. Pattern of infection in invaded households and relation of age and prior antibody to occurrence of infection and related illness. *Am J Epidemiol* 1982; **116**: 228–242.

10. Smith DJ, Lapedes AS, de Jong JC *et al.* Mapping the antigenic and genetic evolution of influenza virus. *Science* 2004; **305**: 371–376.

11. Webster RG, Peiris M, Chen H *et al.* H5N1 outbreaks and enzootic influenza. *Emerg Inf Dis* 2006; **12**: 3–8.

12. Webster RG, Bean WJ, Gorman OT *et al.* Evolution and ecology of influenza A viruses. *Microbiol Rev* 1992; **56**: 152–179.

13. Alexander DJ. A review of avian influenza viruses in different bird species. *Vet Microbiol* 2000; **74**: 3–13.

14. Webster RG, Laver WG. Pandemic variation of influenza viruses. In: Kilbourne ED, editor. *The Influenza Viruses and Influenza.* Academic Press, 1975; pp. 269–314.

15. Shortridge KF. Pandemic influenza: a zoonosis? *Semin Respir Infect* 1992; **7**: 11–25.

16. Webster RG. Influenza virus: transmission between species and relevance to emergence of the next human pandemic. *Arch Virol Suppl* 1997; **13**: 105–113.

17. Horimoto T, Kawaoka Y. Pandemic threat posed by avian influenza A viruses. *Clin Microbiol Rev* 2001; **14**: 129–149.

18. De Jong JC, Claas ECJ, Osterhaus ADME, Webster RG, Lim WL. A pandemic warning? *Nature* 1997; **389**: 554.

19. Claas EC, Osterhaus AD, van Beek R *et al.* Human influenza A H5N1 virus related to a highly pathogenic avian influenza virus. *Lancet* 1998; **351**: 472–477.

20. Subbarao K, Klimov A, Katz J *et al.* Characterisation of an avian influenza A (H5N1) virus isolated from a child with a fatal respiratory disease. *Science* 1998; **279**: 393–396.

21. Guan Y, Shortridge KF, Krauss S, Webster RG. Molecular characterization of H9N2 influenza viruses: were they the donors of the "internal" genes of H5N1 viruses in Hong Kong? *Proc Natl Acad Sci USA* 1999; **96**: 9363–9367.

22. Webster RG. 1918 Spanish influenza: the secrets remain elusive. *Proc Natl Acad Sci* 1999; **96**: 1164–1166.

23. Reid AH, Fanning TG, Hultin JV *et al.* Origin and evolution of the 1918 "Spanish" influenza virus hemagglutinin gene. *Proc Natl Acad Sci* 1999; **96**: 1651–1656.

24. Gamblin SJ, Haire LF, Russell RJ *et al.* The structure and receptor binding properties of the 1918 influenza hemagglutinin. *Science* 2004; **303**: 1838–1842.

25. Taubenberger JK, Reid AH, Lourens RM *et al.* Characterization of the 1918 influenza virus polymerase genes. *Nature* 2005; **437**: 889–893.

26. Bean WJ, Schell M, Katz J *et al.* Evolution of the H3 influenza virus hemagglutinin from human and nonhuman hosts. *J Virol* 1992; **66**: 1129–1138.

27. Ito T, Couceiro JN, Kelm S *et al.* Molecular basis for the generation in pigs of influenza A viruses with pandemic potential. *J Virol* 1998; **72**: 7367–7373.

28. Nakajima K, Desselberger U, Palese P. Recent human influenza A (H1N1) viruses are closely related genetically to strains isolated in 1950. *Nature* 1978; **274**: 334–339.

29. Potter CW. Chronicle of influenza pandemics. In: Nicholson KG, Webster RG, Hay AJ, editors. *Textbook of Influenza*. Oxford: Blackwell Science, 1998; pp. 3–18.

30. Beveridge WIB. The chronicle of influenza epidemics. *Hist Phil Life Sci* 1991; **13**: 223–235.

31. Tumpey TM, Basler CF, Aguilar PV *et al.* Characterization of the reconstructed 1918 Spanish influenza pandemic virus. *Science* 2005; **310**: 77–80.

32. Guan Y, Shortridge KF, Krauss S *et al.* H9N2 influenza viruses possessing H5N1-like internal genomes continue to circulate in poultry in southeastern China. *J Virol* 2000; **74**: 9372–9380.

33. Guan Y, Peiris JSM, Lipatov AS *et al.* Emergence of multiple genotypes of H5N1 avian influenza viruses in Hong Kong SAR. *Proc Natl Acad Sci USA* 2002; **99**: 8950–8955.

34. Apisarnthanarak A, Kitphati R, Thongphubeth K *et al.* Atypical avian influenza (H5N1). *Emerg Infect Dis* 2004; **10**: 1321–1324.

35. de Jong MD, Bach VC, Phan TQ *et al.* Fatal avian influenza A (H5N1) in a child presenting with diarrhea followed by coma. *New Engl J Med* 2005; **352**: 686–691.

36. Tran TH, Nguyen TL, Nguyen TD *et al.* Avian influenza A (H5N1) in 10 patients in Vietnam. *New Engl J Med* 2004; **350**: 1179–1188.

37. World Health Organization. Cumulative number of confirmed human cases of avian influenza A/(H5N1) reported to WHO. (www.who.int/csr/disease/avian_influenza/country/en/index.html)

38. Peiris M, Yuen KY, Leung CW *et al.* Human infection with influenza H9N2. *Lancet* 1999; **354**: 916–917.

39. Fouchier RA, Schneeberger PM, Rozendaal FW *et al.* Avian influenza A virus (H7N7) associated with human conjunctivitis and a fatal case of acute respiratory distress syndrome. *Proc Natl Acad Sci USA* 2004; **101**: 1356–1361.

40. World Health Organization. Influenza pandemic preparedness plan. The role of WHO and guidelines for national or regional planning,1999. (www.who.int/csr/resources/publications/influenza/WHO_CDS_CSR_EDC_99_1/en/)

41. World Health Organization. Global influenza preparedness plan. The role of WHO and recommendations for national measures before and during pandemics, 2005. (www.who.int/csr/resources/publications/influenza/WHO_CDS_CSR_GIP_2005_5/en/index.html)

42. Stöhr K. The global agenda on influenza surveillance and control. *Vaccine* 2003; **21**: 1774–1748.

43. World Health Organization. Global agenda on influenza – adopted version. Part I. *Wkly Epidemiol Rec* 2002; **77**: 179–182; Part II. *Wkly Epidemiol Rec* 2002; **77**: 191–196. (www.who.int/docstore/wer/pdf/2000/wer7535.pdf) (www.who.int/docstore/wer/pdf/2002/wer7728.pdf)

44. World Health Organization. Meeting on avian influenza and human pandemic influenza. Geneva, 7–9 November 2005. (www.who.int/mediacentre/events/2005/avian_influenza/en/index.html)

45. International pledging conference on avian and human pandemic influenza. Co-sponsored by the Government of the People's Republic of China, the European Commission & the World Bank. Beijing, 17–18 January 2006. (www.who.int/dg/lee/speeches/2006/flumeeting_beijing/en/index.html)

46. World Health Organization. Guidelines on the Use of Vaccines and Antivirals during Influenza Pandemics. Geneva, 2004. Document WHO/CDS/CSR/RMD/2004.8. (www.who.int/csr/resources/publications/influenza/en/11_29_01_A.pdf)

47.World Health Organization. Meeting on development and evaluation of influenza pandemic vaccines. Geneva, 2–3 November 2005.
(www.who.int/vaccine_research/diseases/influenza/mtg_021205/en/)

48. European Medicines Agency (EMEA). Guideline on submission of marketing authorisation application for pandemic influenza vaccines through the centralised procedure. April 2004. Document EMEA/CPMP/VEG/4986/03.
(www.emea.eu.int/pdfs/human/vwp/498603en.pdf)
(www.emea.eu.int/pdfs/human/vwp/471703en.pdf)

49. Health Protection Agency. Health Protection Agency coordinates EU pandemic influenza exercise.
(www.medicalnewstoday.com/medicalnews.php?newsid=32376)

4. The Immune Response to Influenza Infection

When the influenza virus infects the cells of the respiratory tract, both innate and adaptive immune responses are stimulated. The innate immune response develops very quickly and controls virus replication during the early stages of infection. While the innate immune system recognizes virus-infected cells through mechanisms that are not antigen-specific, the cytokines produced during this early phase of the host's defence do facilitate activation of subsequent antigen-specific adaptive immune mechanisms.[1] An important element of the transition from the innate to the adaptive immune response is the stimulation of Toll-like receptors (TLRs) in endosomes of antigen-presenting cells (primarily dendritic cells).[2] TLRs recognize and bind to structural components such as single-stranded viral RNA, which are shared by different pathogens, and are important triggers of the "danger signal". Stimulation of immunological memory from prior exposure to viral antigens (by natural infection with the virus or through vaccination) also stimulates specific pathways in the adaptive immune response. Stimulation of immunological memory accelerates the adaptive response, which is delayed during a primary exposure to the antigen. A major aspect of the adaptive immune response involves virus binding to immunoglobulin receptors on B lymphocytes, which subsequently differentiate to plasma cells that produce virus-specific antibodies.[3] Activation of the adaptive immune response occurs through peptides derived from viral proteins, which are presented on antigen-presenting cells to the T lymphocytes. Helper T cells, through the production of cytokines, contribute to B cell proliferation and differentiation to plasma cells, and to the activation and proliferation of virus-specific cytotoxic T lymphocytes (CTLs).[4] Figure 16 details many of the interactions of the different cell populations involved.

Key Messages
• Innate immune responses control influenza virus replication during the early stages of infection.
• Adaptive immune responses are antigen-specific and develop immunological memory that provides a more rapid response upon re-exposure to the virus.
• Adaptive immune response is activated by antigen presentation through professional antigen-presenting cells, dendritic cells (DCs) in particular, which stimulate T cells.
• T helper type 1 (Th1) responses stimulate antibody production (IgG2a) and cytotoxic T lymphocytes (CTLs). T helper type 2 (Th2) responses stimulate antibody production (IgG1), but not CTLs.
• In the effector phase of the adaptive immune response, secretory antibodies prevent infection at mucosal surfaces of the respiratory tract, while circulating antibodies diffuse to and protect the lungs.
• Cytotoxic T lymphocytes (CTLs) contribute to elimination of the infection by lysing virus-infected cells.
• Immunological memory for B cell responses is lifelong, and subtype- and strain-specific. In contrast, T cell memory is more cross-reactive among different subtypes of influenza, while the duration of T cell memory is variable, even among younger adults (months to years).
• Since antibodies are strain-specific, influenza vaccine composition must be updated regularly for protection against newly circulating strains of the virus.
• A significant decline in CTL-mediated immunity is directly linked to the increased risk of influenza illness in older people.

Virus-specific antibodies and CTLs represent the principal effector mechanisms of the adaptive immune response. Influenza virus strain-specific antibodies bind to and neutralize the virus on mucosal surfaces to prevent entry and replication inside the cell. Both IgA (secretory antibodies) and IgG (plasma antibodies) contribute to strain-specific protection along the respiratory tract. T helper (Th1 and Th2) cytokines reciprocally down-regulate each other and only Th1 cytokines facilitate CTL activation.[5] Activation of CTLs and the associated increase in granzyme B (Grz B) that leads to programmed cell death (apoptosis) of virus-infected cells is another critical component of the adaptive antiviral immune response.[6,7] With ageing, the loss of CTL-mediated immunity has been directly linked to the increased risk of influenza illness in older people, while antibody responses appear to be relatively preserved in otherwise healthy older adults.

This chapter will first briefly discuss innate immunity to viral infections and then focus on the adaptive immune response to influenza with special reference to how this response is affected during ageing.

The innate immune response

Key features of the innate response

The "innate immune response" is stimulated when cells are infected with viral, bacterial or fungal pathogens. The influenza virus induces chemokine and cytokine production by infected epithelial cells and monocytes/macrophages. These chemokines attract immune cells to the site of infection, including macrophages, neutrophils and natural killer (NK) cells, as summarized in Table 5. These in turn produce additional cytokines, chemokines and other antiviral proteins.[8] This process activates a number of immune cells, including those of the adaptive immune response.

Type I interferons (IFN-α/β) are among the most important cytokines produced by the innate immune response and have several important antiviral functions. Firstly, they bind to neighbouring cells and induce an antiviral state by promoting the production of several intracellular antiviral proteins that inhibit protein synthesis (including, of course, viral proteins). Secondly, they recruit monocytes/macrophages, T cells and NK cells. Thirdly. they enhance maturation of antigen-presenting cells (APCs)

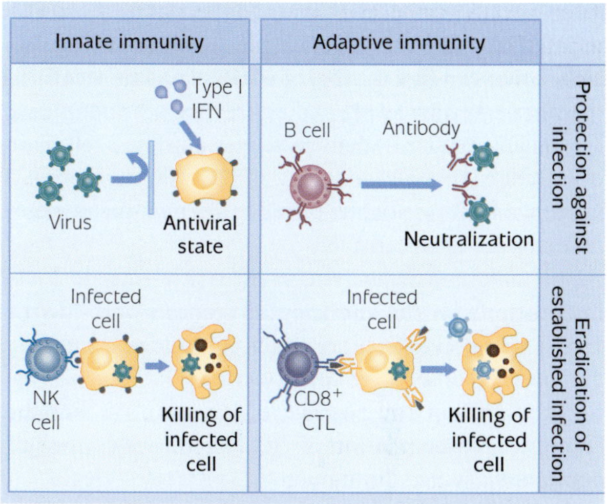

Figure 16. Immune responses to viral infections. Innate immune mechanisms include the production of cytokines and the activation of antigen-presenting cells (macrophages and dendritic cells) and natural killer (NK) cells. The adaptive, antigen-specific, immune response involves both the generation of antibodies by B cells and the activation of both T helper cells (mainly CD4-positive T cells) and CTLs (mainly CD8-positive T cells). Antibodies restrict the spread of virus primarily by neutralizing the virus so that it cannot enter the cell and replicate. CTLs recognize and destroy virus-infected cells, shortening the duration of viral shedding, and are responsible for clearing virus-infected cells from the lungs. Reproduced from Abbas AK, Lichtman AH. *Cellular and Molecular Immunology*, 5th edn. WB Saunders, 2003 with permission from Elsevier.

and increase the expression of major histocompatibility complex (MHC) class I and II molecules on APCs, resulting in enhanced antigen presentation, which facilitates adaptive immune mechanisms.

In addition to cytokines, NK cells play an important role in the innate immune response against viral infections. NK cells are large granular lymphocytes that detect virus-infected cells in a non-specific manner. Many viruses induce a down-regulation of MHC class I molecules on the surface of infected cells, so as to escape destruction by CTLs. NK cells, however, sense the loss of MHC class I molecules on the surface of these cells, and destroy them by apoptosis.[9] The innate immune response is activated within a few hours of infection and lasts for as long as 7 days during a primary influenza infection, when the adaptive immune response takes longer to be activated.

Implications for the immunopathogenesis of influenza
High levels of cytokines are produced in the inflammatory, antiviral response to the large quantities of virus that may be produced in the absence of an adaptive immune response. Desquamation of the respiratory epithelium combined with inflammatory processes leads to

Components of the innate immune response	
	Innate system
Cellular components	Monocytes/macrophages Neutrophils Eosinophils Basophils Mast cells Natural killer cells
Secreted components	Complement Cytokines Lysozyme Acute phase proteins Interferons

Table 5. Components of the innate immune response. Reproduced from Griffin J *et al*. *Crash Course: Immunology and Haematology*, 2003 with permission from Elsevier.

transudation of large amounts of fluid into the lungs. Progressive hypoxia and acute respiratory distress syndrome may cause death within the first 1–2 days of the onset of illness. This process is more likely to occur with pandemic influenza where the individual has not been exposed to the new virus subtype before and must rely on the innate immune response to decrease viral replication. The adaptive immune system must mount a primary response, which is delayed because of a lack of prior exposure to the new subtype of influenza virus.

Recently it has been shown in mice that the lethality of the 1918 pandemic influenza virus is related to high levels of virus replication, severe lung inflammation and infiltration of the lungs with neutrophils and alveolar macrophages.[10,11] These elements are the result of an ineffective innate immune response to control replication of this subtype of influenza in the absence of an adaptive immune response. The limited efficacy of the innate immune response against the 1918 virus appears to be related to adaptation of the NS1 gene to efficiently suppress the IFN-α/β system, thus allowing an almost unrestricted replication of the virus.

Initiation of the adaptive immune response

Interferons, macrophages and NK cells slow virus replication and prevent the spread of virus during the first few days of infection. However, as is seen in individuals whose immune systems are naive to the infecting strain of virus, such as in infants or with pandemic strains of influenza, the innate immune response is not enough to stem the spread of viral infection. Thus, survival relies on the "adaptive" immune response that is initiated during this period and is essential to eliminate the virus completely. This adaptive response involves the activation of both bone marrow-derived lymphocytes (B cells) and thymus-derived lymphocytes (T cells), resulting in humoral and cell-mediated immunity, respectively. Virus-activated T cells, through a variety of cytokine mediators, stimulate B cells to differentiate and produce antibodies that are specific for the strain of virus involved.

A key feature of the adaptive immune response is the ability to store "immunological memory" for a specific pathogen. This memory provides a more rapid adaptive immune response that is specifically targeted to the antigens contained in the pathogen. Immunological memory results from differentiation of virus-specific B and T cells that are programmed to respond to specific epitopes, in this case of the viral antigens. Vaccination targets the adaptive immune response to stimulate B cells to differentiate to plasma cells that produce circulating antibodies and an army of virus-specific T cells that are activated upon re-exposure to the pathogen during a natural infection.

Structure and function of MHC

The adaptive immune response against influenza is activated when T cells recognize viral peptides presented by the major histocompatibility complex (MHC). The genetic locus that codes for MHC is a cluster of tightly linked genes, found on the short arm of chromosome 6. The MHC complex is involved in peptide binding, processing and presentation. It mediates presentation of a vast array of peptides derived from different pathogens. MHC molecules also allow the immune system to distinguish "self" from "non-self" and in this context are referred to as "transplantation antigens". MHC genes exhibit a high degree of polymorphism and, within the individual genotype, have the capacity to produce MHC molecules that bind to a specific subset of pathogen-derived peptides determined by the MHC molecules of the individual. Thus, each individual exhibits a unique set of MHC molecules and the diversity of the MHC increases the likelihood that a protective adaptive immune response can be mounted by most people.

Class I and class II MHC molecules are glycoproteins expressed on the cell surface, and consist of cytoplasmic, transmembrane and extracellular portions. The overall structure of individual MHC molecules enables the immune system to distinguish self from non-self. Indeed, on this basis T cells are able to recognize non-self MHC molecules – for example, in the case of rejection after organ transplantation. The pathogen specificity of MHC is largely concentrated in the peptide-binding cleft of the molecules, where peptides from the pathogen are presented to T cells. These T cells are only able to recognize these peptide antigens when they are presented by self-MHC molecules (self-MHC restriction). CD8-positive T cells recognize antigen only in association with class I MHC molecules (class I MHC-restricted), while CD4-positive T cells recognize antigen only in association with class II MHC molecules (class II MHC-restricted).

Antigen processing and presentation

MHC molecules present peptide fragments derived from viral proteins either degraded or synthesized inside the cell, and the MHC–peptide complex is transported to the cell surface, where the complex can be recognized by the T cell receptor.[12] Antigen-presenting cells (APCs) have different pathways of antigen processing for class I and class II MHC, and different requirements for loading peptides from live or killed virus onto the MHC class I molecules (see Figure 17). In contrast to other APCs, dendritic cells (DCs) are able to present peptides derived from any form of virus, including killed viruses, through what have been defined as "non-classical" pathways of antigen processing and presentation. The ability of DCs to deliver viral protein antigens from the endosomal compartment to the cell cytosol for processing in the MHC class I pathway is important in the context of the development of new inactivated influenza vaccines that will induce not only an adequate antibody response but also CTL activity, as discussed further in Chapter 9.

Professional APCs process and present antigen to CD4-positive T cells (T helper cells) in association with MHC II, and to CD8-positive T cells (CTLs) in association with

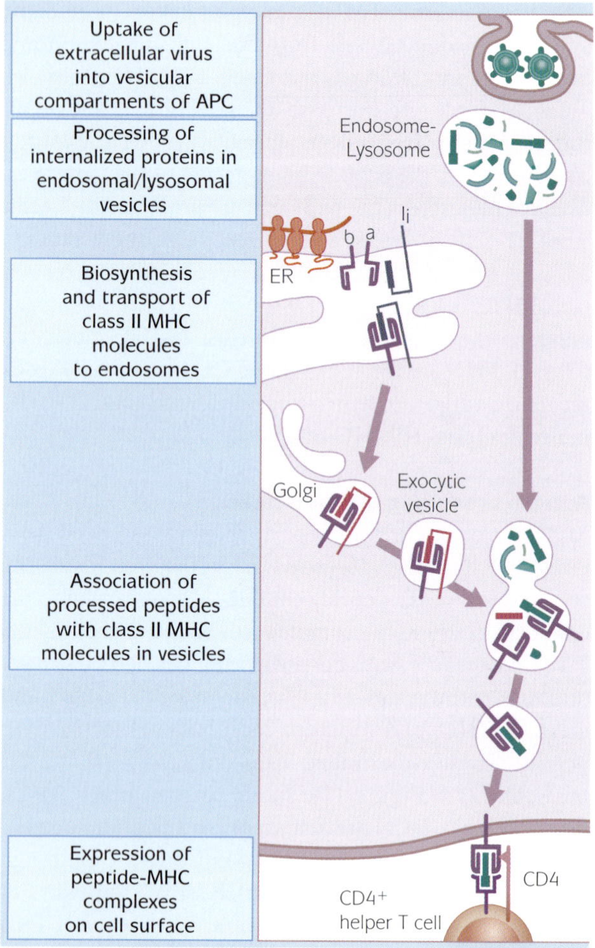

Figure 17. Pathways of antigen processing and presentation. Peptides presented on class II MHC are processed from antigens taken up by endocytosis by the antigen-presenting cell and loaded onto MHC II for presentation to the CD4-positive (CD4$^+$; helper) T cells. Peptides presented on class I MHC are derived from cytosolic proteins degraded by proteasomes, the resulting peptides being transported to the endoplasmic reticulum (ER) for loading onto MHC I molecule for subsequent presentation on

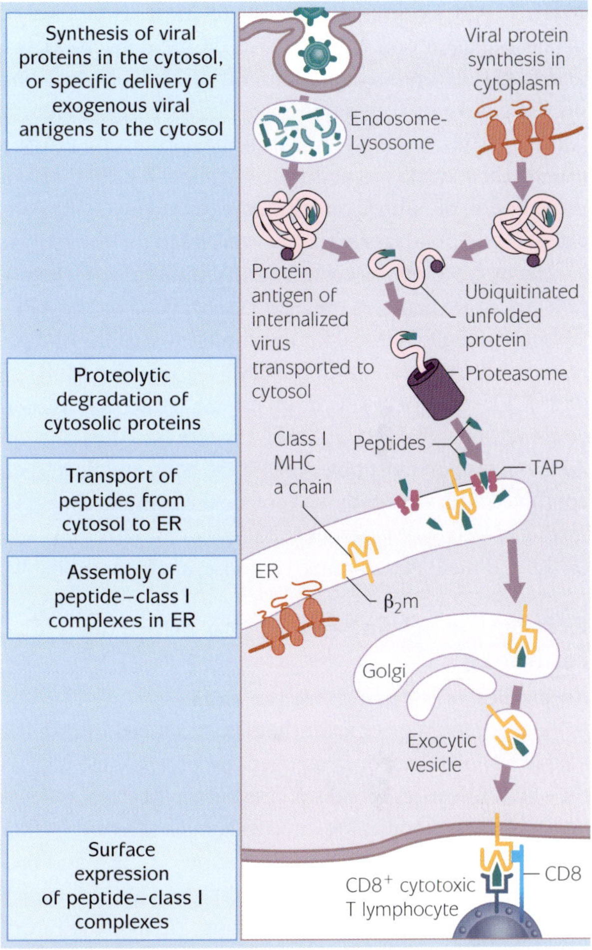

the cell surface to CD8-positive (CD8⁺; cytotoxic) T cells. Antigen-presenting cells present viral peptides derived from replicating virus inside the cell. Dendritic cells also have the unique capability to present peptides derived from exogenous virus taken up by the cell. Adapted from Abbas AK, Lichtman AH. *Basic Immunology*, 2nd edn. WB Saunders, 2004 with permission from Elsevier.

MHC I. APCs express high levels of MHC II (expressed almost exclusively on APCs, such as DCs and macrophages, and on B cells) and variable amounts of MHC I depending on the degree of upregulation of MHC I by interferon-gamma (IFN-γ). DCs represent the key professional APCs due to their exquisite ability to activate T cells through presentation of antigen on MHC molecules (involving both classical and non-classical pathways) and to further activate T cells by a variety of costimulatory molecules. Helper T cells recognize peptides presented by MHC II on the APC. This interaction is critical for the influenza-specific T helper cell responses necessary for the development of both B cell and CTL memory. CTLs recognize viral peptides in conjunction with MHC I; MHC I is expressed not only on APC but also on most other cells in the body and thus when combined with viral peptides provide the stimulus for CTL-mediated killing of virus-infected cells in the effector phase of the cell-mediated immune response, as discussed below.

Effector mechanisms of the adaptive response

Antibodies to HA neutralize the virus

Influenza strain-specific antibodies, produced during the adaptive immune response, represent the principal mechanism by which infection of cells is prevented. The main virus-neutralizing antibodies are directed against the viral surface glycoprotein HA (see Figure 7 in Chapter 2). IgA and IgG are the main immunoglobulin subtypes that neutralize influenza virus on the mucosal surface of the respiratory tract. IgA is specifically designed for secretion onto the mucosal surface where these secretory antibodies provide the first line of defence, while IgG subtype antibodies circulate in the blood and transudate to the airways and lungs. Binding of anti-HA antibodies blocks the receptor-binding site on the HA, as shown in Figure 18. This prevents attachment and cell entry of the virus particles, thus effectively neutralizing the virus. Furthermore, NK cells bind to antibody–virus complexes on the surface of infected cells and

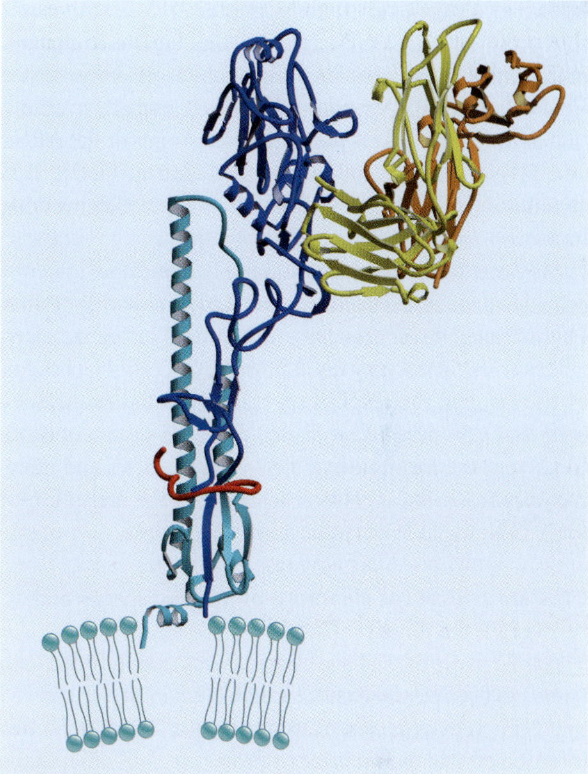

Figure 18. Mechanism of influenza virus neutralization by an antibody directed against the viral HA. The figure shows an HA monomer (in blue), with (in yellow) an attached Fab (antigen-binding) fragment of an antibody specific for epitope D of HA (see Figure 7). This figure was produced as described in the caption to Figure 6, on the basis of the co-ordinate file from the Protein Data Bank, code 1EO8. Fleury D *et al*. Structural evidence for recognition of a single epitope by two distinct antibodies. *Proteins Struct Funct Genet* 2000; **40**: 572.

kill by antibody-dependent cell-mediated cytotoxicity (ADCC) as part of the adaptive immune response.

The peptide sequences of the antigenic epitopes of HA (and NA) change as a result of selective pressure by the immune system against the native virus, a phenomenon known as antigenic drift (see Chapter 3). Mutant viruses thus produced may escape antibody binding due to changes in the antigenic epitopes determining the B cell response (see Figure 7). Influenza vaccines are updated annually to ensure that immunity to the relevant predicted strains of the H3N2 and H1N1 subtypes of influenza A and influenza B is stimulated (see Chapters 3 and 8). Thus, antibodies resulting from exposure to a different strain of influenza that circulated in previous years or was contained in a vaccine, may not provide adequate protection against the circulating strains. In this case, antibodies may not control influenza virus infection and illness may result from the successful replication of the virus in the respiratory tract. Infected respiratory epithelial cells need to be cleared from the system in order to prevent the formation of new virus particles and allow recovery from illness. This is achieved by the activation of the T cell-mediated adaptive immune response.

CTLs are critical for clearance of influenza virus and recovery from illness

The importance of CTLs in clearing virus from infected lung tissue has been clearly demonstrated. Human studies showed that CTL activity is important for recovery from influenza infection even in the presence of protective antibodies to the infecting virus strain.[13] CTLs combat influenza viral infections by recognizing and destroying virus-infected host cells that otherwise become the factories for viral replication. Infected cells expressing on their surfaces the MHC I–viral peptide complex are recognized by virus-specific CTLs. The mechanisms by which CTLs induce lysis of virus-infected cells include perforin- or granule-mediated killing.[4] Granule-mediated killing appears to be the critical effector function of CTLs against influenza-infected cells.

Virus-specific, granule-mediated killing results from the production of granules within CTLs that contain pore-forming proteins (perforins) and granzymes (Figure 19). Granules are transported to the surface of activated CTLs at the point of contact with an appropriate target cell.

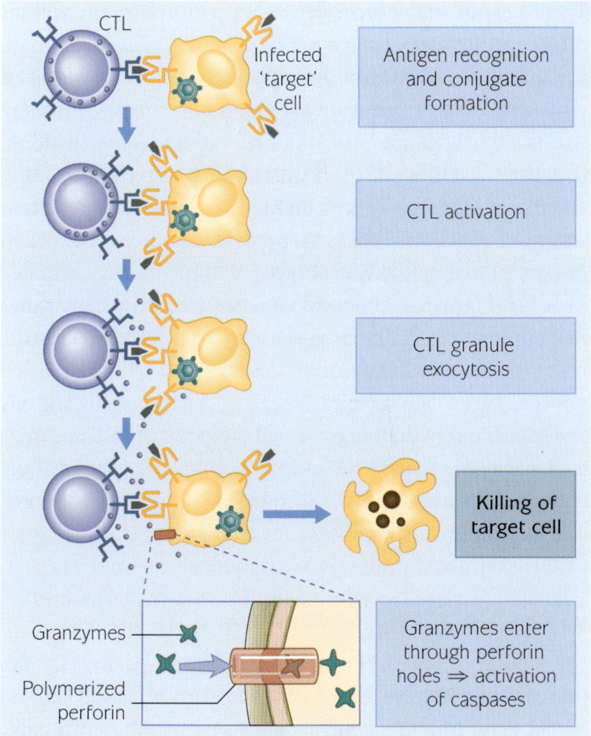

Figure 19. Mechanism of CTL-mediated killing. Granules in the CTL migrate to the cell surface upon CTL activation and cross into the virus-infected cell to induce apoptotic cell death through DNA fragmentation and disruption of mitochondrial metabolism. Reproduced from Abbas AK, Lichtman AH. *Basic Immunology*, 2nd edn. WB Saunders, 2004 with permission from Elsevier.

Through a process facilitated by perforin, granzymes are transported across the cell membrane into the target cell and subsequently trigger an enzymatic cascade that eventually results in DNA fragmentation and cell death (apoptosis). The combined activity of Grz B and perforin is important for target-cell lysis. Thus, the activation of CTLs and associated increase in Grz B that leads to apoptotic cell death is a critical component of antiviral immunity.

T cell responses offer more cross-protection between different strains of influenza

In contrast to the antibody response to a specific strain of influenza virus, T cell responses tend to be cross-protective within the types and subtypes of influenza, as the viral peptides that stimulate the T cell response are relatively more conserved than antibody binding sites across the different strains of influenza.[14] Internal peptide sequences of HA and NA are similar within the subtypes of influenza (e.g. A/H3N2 vs A/H1N1) and the internal viral proteins – the matrix and nucleoproteins (NP) – are relatively conserved within the types of influenza (e.g. influenza A vs B). Although T cell responses are much more cross-reactive compared to antibody responses, recent data suggest that antigenic drift also affects CTL responses against NP as soon as immunological pressure is applied.[15] Although B cell memory is largely strain-specific, it lasts a lifetime. In contrast, T cell memory is more cross-reactive but the duration is much more short-lived.[16]

Effect of ageing on the immune system

Immunosenescence is the global term for age-related changes that are associated with an increased risk of infection in older adults. A gradual shift with ageing from predominantly naive T cells to increasing proportions of memory T cells reflects the multiple immune responses that are mounted throughout life. Thymic involution and the loss of naive T cells with ageing will ultimately exhaust the capacity of the immune system to respond to new antigens and may thus result in an increased risk of contracting infections, including influenza, as a hallmark of immunosenescence. Recent studies suggest that although early immune responses to influenza may be similar in young and older adults, the ability to maintain or expand memory helper T cells that can respond to influenza challenge declines with ageing.[17] Although mucosal and systemic antibodies are the first line of defence of the adaptive immune response, just increased levels of mucosal antibodies to influenza do not necessarily translate into clinically meaningful differences in protection against influenza.

Costimulatory molecules, the regulation of cytokine production by both helper T cells and CTLs, and cytolytic mediators from both natural killer cells and CTLs, also have critical roles in the response to influenza infection and recovery from illness. The age-related loss of CD28 expression on CTLs is associated with replicative senescence, growth arrest, increased production of several pro-inflammatory cytokines, resistance to apoptosis,[18] and the loss of CD28-mediated costimulation required for optimal stimulation of CTLs.[19,20] CTLs that lack CD28 expression have little or no cytolytic activity.[21] Lack of CD28 expression on T cells is also associated with a decline in antibody responses to influenza vaccination.[22,23] Although the frequency of influenza virus-specific CTLs does not appreciably change with age,[24] the decline in CTL activity is associated with decreases in influenza-specific CTL responses to influenza challenge[25] and the production of cytolytic mediators including GrzB.[26] In addition to higher influenza attack rates in the population of 65 years and older, these changes in CTL function correspond to the related increased risk of serious complications that require hospitalization, prolong recovery from illness and may lead to death in older people (see also Chapter 5).

References

1. Durbin JE, Fernandez-Sesma A, Lee CK *et al.* Type I IFN modulates innate and specific antiviral immunity. *J Immunol* 2000; **164**: 4220–4228.

2. Diebold SS, Kaisho T, Hemmi H *et al.* Innate antiviral responses by means of TLR7-mediated recognition of single-stranded RNA. *Science* 2004; **303**: 1529–1531.

3. Lee BO, Rangel-Moreno J, Moyron-Quiroz JE *et al.* CD4 T cell-independent antibody response promotes resolution of primary influenza infection and helps to prevent reinfection. *J Immunol* 2005; **175**: 5827–5838.

4. Doherty PC, Topham DJ, Tripp RA *et al.* Effector CD4+ and CD8+ T-cell mechanisms in the control of respiratory virus infections. *Immunol Rev* 1997; **159**: 105–117.

5. Doherty PC, Allan W, Eichelberger M, Carding SR. Roles of alpha beta and gamma delta T cell subsets in viral immunity. *Ann Rev Immunol* 1992; **10**: 123–151.

6. Johnson BJ, Costelloe EO, Fitzpatrick DR *et al.* Single-cell perforin and granzyme expression reveals the anatomical localization of effector CD8+ T cells in influenza virus-infected mice. *Proc Natl Acad Sci USA* 2003; **100**: 2657–2662.

7. Lawrence CW, Ream RM, Braciale TJ. Frequency, specificity, and sites of expansion of CD8+ T cells during primary pulmonary influenza virus infection. *J Immunol* 2005; **174**: 5332–5340.

8. Julkunen I, Melen K, Nyqvist M, Pirhonen J, Sareneva T, Matikainen S. Inflammatory responses in influenza A virus infection. *Vaccine* 2000; **19**(Suppl 1): S32–S37.

9. Trapani JA, Smyth MJ. Killing by cytotoxic T cells and natural killer cells: multiple granule serine proteases as initiators of DNA fragmentation. *Immunol Cell Biol* 1993; **71**(Pt 3): 201–208.

10. Tumpey TM, Basler CF, Aguilar PV *et al.* Characterization of the reconstructed 1918 Spanish influenza pandemic virus. *Science* 2005; **310**: 77–80.

11. Tumpey TM, Garcia-Sastre A, Taubenberger JK *et al.* Pathogenicity of influenza viruses with genes from the 1918 pandemic virus: functional roles of alveolar macrophages and neutrophils in limiting virus replication and mortality in mice. *J Virol* 2005; **79**: 14933–14944.

12. Braciale TJ, Morrison LA, Sweetser MT *et al.* Antigen presentation pathways to class I and class II MHC-restricted T lymphocytes. *Immunol Rev* 1987; **98**: 95–114.

13. McMichael AJ, Gotch FM, Noble GR, Beare PA. Cytotoxic T-cell immunity to influenza. *New Engl J Med* 1983; **309**: 13–17.

14. Thomas PG, Keating R, Hulse-Post DJ, Doherty PC. Cell-mediated protection in influenza infection. *Emerg Infect Dis* 2006; **12**: 48–54.

15. Rimmelzwaan GF, Boon AC, Voeten JT *et al.* Sequence variation in the influenza A virus nucleoprotein associated with escape from cytotoxic T lymphocytes. *Virus Res* 2004; **103**: 97–100.

16. Ennis FA, Yi-Hua Q, Schild GC. Antibody and cytotoxic T lymphocyte responses of humans to live and inactivated influenza vaccines. *J Gen Virol* 1982; **58**: 273–281.

17. Kang I, Hong MS, Nolasco H *et al.* Age-associated change in the frequency of memory CD4+ T cells impairs long term CD4+ T cell responses to influenza vaccine. *J Immunol* 2004; **173**: 673–681.

18. Effros RB, Dagarag M, Valenzuela HF. *In vitro* senescence of immune cells. *Exp Gerontol* 2003; **38**: 1243–1249.

19. Effros RB, Boucher N, Porter V *et al.* Decline in CD28+ T cells in centenarians and in long-term T cell cultures: a possible cause for both *in vivo* and *in vitro* immunosenescence. *Exp Gerontol* 1994; **29**: 601–609.

20. Engwerda CR, Handwerger BS, Fox BS. Aged T cells are hyporesponsive to costimulation mediated by CD28. *J Immunol* 1994; **152**: 3740–3747.

21. Effros RB. Genetic alterations in the ageing immune system: impact on infection and cancer. *Mech Ageing Dev* 2003; **124**: 71–77.

22. Goronzy JJ, Fulbright JW, Crowson CS *et al.* Value of immunological markers in predicting responsiveness to influenza vaccination in elderly individuals. *J Virol* 2001; **75**: 12182–12187.

23. Saurwein-Teissl M, Lung TL, Marx F *et al.* Lack of antibody production following immunization in old age: association with CD8(+)CD28(-) T cell clonal expansions and an imbalance in the production of Th1 and Th2 cytokines. *J Immunol* 2002; **168**: 5893–5899.

24. Boon AC, Fringuelli E, Graus YM *et al.* Influenza A virus specific T cell immunity in humans during aging. *Virology* 2002; **299**: 100–108.

25. McElhaney JE, Xie D, Hager WD *et al.* T cell responses are better correlates of vaccine protection in the elderly. *J Immunol* 2006, in press.

26. McElhaney JE. The unmet need in the elderly: designing new influenza vaccines for older adults. *Vaccine* 2005; **23** (Suppl 1): S10–S25.

5. Pathogenesis, Clinical Features and Diagnosis

During an influenza illness, the primary causes of disease are related to infection and replication of the virus in the respiratory epithelium. These cellular "factories" for virus production ultimately lead to lysis of the epithelial cell and desquamation of the respiratory lining. At the same time, sensors of viral replication stimulate the production of proinflammatory cytokines, including IL-1, IL-6 and TNF-α, which circulate in the body and are responsible for the systemic symptoms associated with influenza illness, such as fever, muscle aches and malaise.

Generally, flu is a self-limiting respiratory infection. The common circulating strains of influenza virus normally remain restricted to the respiratory tract and escape only under exceptional circumstances. As a result, the virus is rarely found circulating in the blood or in other organs. The main reason is that the protease required for cleavage of the viral HA is restricted to the epithelium of the airways and lungs (see Chapter 2). Host factors such as interferons and other components of the immune response may also be important for preventing the spread of the virus outside of the respiratory tract.

When a new influenza A subtype is introduced into the human population, the situation is different. The severity of disease may be considerably increased because of the complete lack of immunity in infected individuals under these conditions. Additional alterations in viral proteins may have contributed to the extremely high pathogenicity of the H1N1 Spanish flu pandemic virus of 1918 and that of the more recent avian H5N1 virus, which, in ongoing outbreaks thus far (March 1, 2006), has infected 179 people with a very high case fatality rate.[1] A hallmark of virulence of highly pathogenic avian influenza viruses, such as the H5 subtype viruses, is the presence of the multibasic cleavage site in the HA molecule, as discussed in Chapter 2. Yet, the 1918 HA does not have the multibasic cleavage site, indicating that other factors contributed to the extreme virulence of the Spanish flu virus. Recent studies, using virus reconstructed from genetic material obtained from victims of the 1918 pandemic, suggest a role for the NS1 protein.

Key Messages

- Influenza spreads by aerosols or droplets – it can persist in the air and on hands or fomites for significant periods of time.
- The pathogenesis of influenza is based on mucosal inflammation and lysis of respiratory epithelial cells.
- Influenza virus does not normally spread beyond the respiratory tract due to the limited tissue distribution of proteases involved in activation of HA.
- Systemic symptoms of influenza are mainly due to circulation of the inflammatory cytokines produced in response to infection.
- Age, co-morbid illness and vaccination status affect the presentation of illness and are key determinants of illness severity.
- Considering that older people may lose 2–3% of muscle power per day of bed rest, significant disability may result from influenza illness.
- Pandemic influenza and the recent – often fatal – cases of human infection with the avian H5N1 virus underscore the high pathogenicity of influenza in immunologically naive individuals.
- Physician awareness is the single most important factor in the diagnostic accuracy of influenza. It is essential therefore that primary care physicians be well informed about local influenza surveillance data.
- Proper and rapid diagnosis is essential to control influenza infections by antiviral treatment and to avoid the inappropriate use of antibiotics.

It appears that this viral protein adapted to antagonize the production of antiviral cytokines that suppress viral replication and are thus important during the early phase of infection. This type of adaptation of the NS1 gene may also be responsible for the enhanced proinflammatory cytokine response to H5N1 infection and the increased lethality of these highly pathogenic viral strains.[2,3]

Influenza is a common illness during the winter months, but the overlap of symptoms with other respiratory illnesses and the practitioner's knowledge of whether influenza is circulating in the community is key to the diagnosis. The severity of illness depends on the age of the patient (the very old and the very young are most susceptible to serious illness), whether or not high-risk conditions (chronic heart, lung and kidney diseases, diabetes or immunosuppressive conditions) are present, and whether or not the person has been vaccinated. Thus, physician knowledge about influenza in the community combined with an understanding of the individual risk of complications of influenza is required for clinical decision-making with respect to vaccination and prescription of antiviral drugs for prevention or treatment of influenza.[4]

Pathogenesis of influenza

The different types of influenza are of variable clinical importance in people. The influenza A H3N2 subtype is usually associated with greater mortality than the H1N1 subtype and the influenza B virus. Influenza C causes clinically unimportant illness. The effects of influenza on the naive immune system are evident in the high-risk population of children under the age of 2 years, with some protection mediated by maternal antibodies in infants less than 6 months old.[5] Elderly people are particularly at risk for influenza illness, because of ageing effects on the immune system and the greater incidence of underlying medical conditions. The more co-morbid conditions, the greater the risk. People in residential care are particularly susceptible because they are generally older, have a high rate of chronic ill health, live in close proximity to each other (facilitating transmission) and respond less well to vaccination. Interestingly, the current cohort of older people appears to be less susceptible compared to young adults and children when H1N1 strains circulate in the community.[6] This protection of older adults against H1N1-mediated illness has been attributed to priming with H1N1 strains that circulated during their childhood and may have generated protective immunological memory.

Mode of transmission

The primary mode of influenza transmission is by aerosol or droplets. The virus may remain suspended in the air for long periods of time and can be dispersed by air currents. Transmission may also occur by contact with virus-contaminated hands or fomites. Inhaled virus may be trapped in mucus produced by airway epithelial cells and then transported by the cilia to the posterior pharynx, where it is swallowed or expectorated ("tracheal toilet"). This mechanism may be impaired in cases of chronic obstructive lung disease or asthma, or in debilitated people.

Lysis of respiratory epithelial cells

The influenza virus causes a lytic infection of respiratory epithelial cells (see Chapter 2). As a consequence, mucosal cells become vacuolated and oedematous, and desquamate, leaving only the basal cells and the basement membrane of the respiratory epithelium. The resulting exudative process with increased mucus production causes a runny nose, cough and nasal congestion. In uncomplicated influenza, acute diffuse mucosal inflammation and oedema of the upper airway and bronchi are observed. In older people, cough is the most common presenting feature of influenza illness, often in the absence of other more classic symptoms such as fever and myalgia, but these symptoms do not distinguish influenza from other viral respiratory illnesses.[7]

With viral pneumonia, there is an interstitial pneumonitis with a predominantly mononuclear leucocyte infiltration.

The alveolar walls become denuded of epithelium, hyaline membranes form, the intra-alveolar space becomes filled with exudate and haemorrhage from the surrounding capillaries, which significantly impairs the diffusion of gases. Progressive hypoxia occurs as the alveolar space is obliterated. In pandemic influenza, where the immune system has no immunological memory for the infecting virus strain, this may be a common scenario even in healthy young people.[4]

Immunopathogenesis

The immune response to influenza has been presented in the previous chapter. This section discusses the immunopathogenesis of influenza – how the body's response to the virus contributes to the symptoms of influenza. Inflammatory cytokines, including tumour necrosis factor-alpha (TNF-α) and type I interferons (IFN-α/β), have been associated with the pathogenesis of influenza infection. Although they are critical components in the down-regulation of intracellular protein synthesis, thus limiting new virus production, these cytokines, along with interleukins (IL-1β, IL-6) produced by the adaptive immune response, are largely responsible for the systemic symptoms of influenza illness, including fever and myalgia. Importantly, influenza virus is limited to the lungs except in rare circumstances where infections of muscles and the central nervous system have been documented. Local symptoms such as cough and sore throat may be a combination of desquamation of the epithelial lining of the airways and the resulting inflammatory response to necrosis of the cells in the process of viral replication. In particular, plasma IL-6 levels correspond to the severity of respiratory symptoms and fever in community-acquired influenza A illness, although serum levels of a number of other cytokines are elevated as well.[8] Paradoxically, older people often do not mount a fever with influenza infection, even though IL-6 levels increase with age. In children, influenza-mediated upper respiratory illness is an important cause of acute otitis media.[9]

> "There was clear evidence of … shortage of breath and the appearance of mahogany spots around the mouth which would extend and coalesce into a violaceous heliotrope cyanosis until 'a white man could not be distinguished from a coloured'. A peculiar stench emanated from many patients. With increasing cyanosis, patients begin to gasp for breath; blood-stained fluid would froth from the mouth; patients would become delirious, and death would follow from suffocation. With rigor mortis, bloody fluid would gush from the mouth and nostrils. The time from hospital admission to death could be a few hours to 2–3 days."
>
> ***Textbook of Influenza* (Ref. 33)**

Pathogenesis of pandemic influenza

In pandemic influenza, a novel virus subtype is introduced into the human population, to which, at that point, no immunity exists (see Chapter 3). Factors contributing to the increased severity of illness under these conditions relate to the induction of a primary viral pneumonia and a profound inflammatory response. Especially in immunologically naive individuals, non-specific destruction of cells by both viral and immunologic mechanisms, and the delay in the activation of the adaptive immune response, are involved in the pathogenesis of the disease. Destruction of respiratory epithelium due to viral replication and the inflammatory exudates produced may lead to profound hypoxia and death within days from the onset of illness, as illustrated in the text box above, describing the typical disease picture caused by the Spanish flu virus.

Extensive sequence analyses of the 1918 virus genome, using RNA material rescued from bodies of victims of the pandemic, as well as reconstruction of viruses containing elements of the 1918 virus, have indicated that adaptation of the NS1 protein may have contributed to the extreme pathogenicity of the virus. The innate immune response to

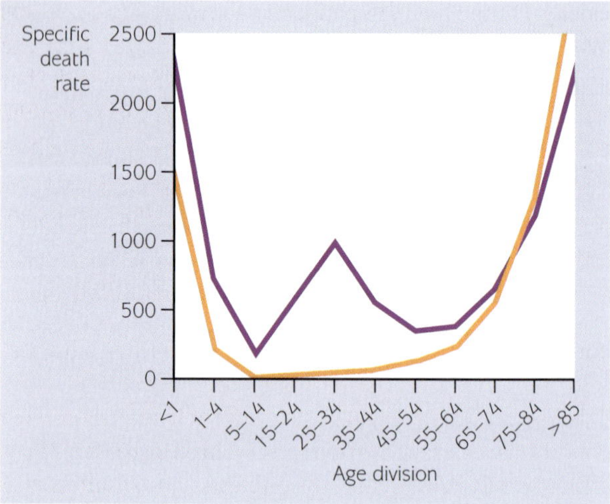

Figure 20. Influenza and pneumonia mortality by age in the USA. Influenza- and pneumonia-specific mortality by age, including the pandemic year 1918 and the average of the interpandemic years 1911–15, is shown. Specific death rate is per 100,000 of the population for each age division. Purple line, 1918 pandemic; orange line, average of interpandemic years 1911–15. Adapted from Potter CW, editor. *Influenza (Perspectives in Medical Virology)*, 2003 with permission from Elsevier.

the influenza includes sensors in the host cell that recognize viral RNA and stimulate the synthesis of type I interferons (IFN-α/β). It appears that in the case of the 1918 pandemic virus, the NS1 gene has adapted to efficiently suppress the IFN-α/β system. Because of the role of type 1 interferons in suppressing viral replication, the inhibitory effect of the 1918 NS1 may have considerably enhanced the pathogenicity of the virus.[2,3] It is likely that other properties of the 1918 virus have contributed to its extreme pathogenicity as well, but these are still not completely understood.[2,3,10,11]

As shown in Figure 20, death was age-related during interpandemic years preceding the Spanish flu outbreak, with high mortality rates in the under 5-year-olds and the elderly. During the 1918 pandemic, the mortality curve was W-shaped, reflecting the higher attack rates that are generally seen in healthy young adults (25–49-year-olds) but with a higher mortality rate in this population during the pandemic compared to interpandemic years. The highest incidence of infection was in 5–14-year-olds, but they had the lowest mortality. Death was frequently due to primary viral pneumonia (usually uncommon) or secondary bacterial pneumonia (antibiotic treatment was not yet available).

Attack rates of Asian flu (1957–58) varied in different communities around the world. Groups with the highest infection rates included school children aged 5–14 years (>50%), the institutionalized and those living in crowded conditions. Overall, mortality was only 10% of that seen with the Spanish flu, but despite this, it translated into 33,000 excess deaths during the second wave in the UK, and 39,000 and 20,000 excess deaths during the first and second waves respectively in the USA. Many factors have been postulated to explain differences in attack rates, but these have yet to be determined (reviewed in Ref. 4).

Presentation of uncomplicated influenza

The clinical presentation of influenza ranges from an asymptomatic infection or a self-limiting upper respiratory tract infection (URTI) to a severe illness, often with serious, potentially fatal, complications.[12] The incubation period is 1–4 days, with an average of 2 days. The clinical course is dependent on the virulence of the virus. The age of the patient (the very young and the old), and the presence of chronic medical illness, such as cardiac and pulmonary diseases, immunosuppression and pregnancy enhance the severity of illness.

The typical uncomplicated influenza A syndrome is a symptom complex that overlaps with a number of other respiratory illnesses. There is an abrupt onset, with high-grade fever (38–40°C) being the most prominent symptom. Other symptoms include headache, chills, dry cough, myalgia, malaise and anorexia. In more extreme cases, severe

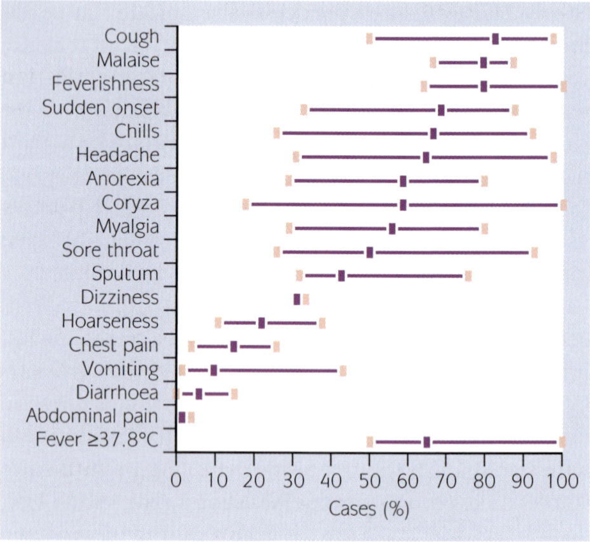

Figure 21. Influenza symptoms and their associated frequencies. Data are from 10 studies involving 520 adults with uncomplicated influenza A. Adapted from Nicholson KG. Human influenza. In: Nicholson KG, Webster RG, Hay AJ, editors. *Textbook of Influenza*. Oxford: Blackwell Science, 1998; pp. 219–264 with permission from Blackwell Publishing.

malaise with prostration is observed. Respiratory symptoms such as rhinorrhoea, nasal congestion and sore throat are present, but are overshadowed by the systemic symptoms during the first 3 days of illness. The influenza virus tends to remain in the respiratory tract and does not usually cause a viraemia. Systemic symptoms are due to proinflammatory cytokines, which are released following activation of the host's defence mechanisms (see Chapter 4). The cough frequently changes from a dry, hacking nature to one that is productive of small amounts of sputum that are usually mucoid but can be purulent. After the fever and upper respiratory tract symptoms resolve (usually within 7–10 days), cough and weakness can persist for 1–2 additional weeks. The frequency of influenza symptoms in adults, shown in Figure 21, may vary with the age of the patient.

Complications of influenza

Influenza-associated pneumonia

In otherwise healthy individuals, influenza infection normally results in an uncomplicated URTI that resolves within 1–2 weeks. However, pneumonia is a relatively common complication (5–38% with influenza A and 10% with influenza B), predominantly in the elderly, patients with chronic cardiopulmonary disease, pregnant women and immunocompromised individuals. The aetiology may be viral, bacterial or mixed viral–bacterial.[12] Such patients can deteriorate rapidly and mortality can be close to 50%.[13]

In primary viral pneumonia, typical influenza is followed by a rapid progression (over 2–3 days) of fever, cough, dyspnoea, chest pain and cyanosis. Physical examination and chest X-ray disclose diffuse bilateral infiltrates consistent with acute respiratory distress syndrome. If fatal, death usually occurs within 4–5 days of first symptoms.

Combined viral–bacterial pneumonia is more common than primary viral pneumonia. Of patients with a severe pneumonia, 75% will have secondary bacterial infection. In these cases, the individual will appear to be recovering from the influenza illness and then have a recurrence of the respiratory symptoms. The bacteria most commonly involved are *Staphylococcus aureus* and *Streptococcus pneumoniae*, with *Haemophilus influenzae* being less common. There is evidence that influenza infection actively facilitates the pathogenicity of bacteria and the impact of illness, causing immunosuppression. In the 1957–58 pandemic, mortality from staphylococcal pneumonia (28%) was similar in all age groups and twice as high as mortality from other pneumonias (12%), which tended to occur in the elderly and those with chronic medical conditions. Once the responsible pathogen has been identified, appropriate antibiotic treatment should be initiated promptly.

Bacterial pneumonia as a complication of influenza also has a different presentation from primary viral pneumonia. Again, patients initially show clinical improvement from illness and then develop worsening respiratory symptoms. In this case, physical and chest X-ray examinations are more likely to show localized signs of consolidation.

Influenza B virus can cause the same spectrum of disease as that seen after influenza A virus infection, and severe illness can occur, particularly in the elderly.

Other respiratory complications

Exacerbations of asthma, chronic obstructive pulmonary disease and cystic fibrosis are common complications of influenza illness. Acute bronchitis develops in 30% of cases and, less commonly, lung abscess and empyema may be observed. Three-quarters of children with asthma will suffer an exacerbation. In hospitalized children, influenza accounts for 16% of asthmatic exacerbations and 4–13% of exacerbations of cystic fibrosis. The most common respiratory complications in hospitalized children are acute bronchitis (12–26%), croup (5–15%) and pneumonia (5–8%).

Non-respiratory complications

Myositis is reported more frequently in children with influenza B, but adults may also be affected and may develop rhabdomyolysis with acute renal failure.[14] Cardiac complications, specifically myocarditis, have been described in patients with influenza A and B, but these complications are mostly asymptomatic.[15] Pericarditis has been reported rarely. ECG abnormalities are common and usually transient (81% of hospitalized patients, 43% of community patients), but underlying conditions can cause fatal arrhythmias or cardiomyopathies. CNS complications are rare and range from irritability and confusion to psychosis and severe encephalopathy due to a variety of inflammatory processes, including Reye's syndrome.[16] Recovery is usually complete. The link between the 1918–20 pandemic and encephalitis lethargica remains unproven. A small increase in the incidence of Guillain–Barré syndrome (GBS) (1–2 excess cases per million against a background of 10–20 cases per million) has been reported after influenza vaccination.[17] However, it is unclear whether this marginal increase is truly associated with influenza vaccination, while, even if it is, the risk is considerably smaller than the risk of death from influenza or its complications, particularly in high-risk groups (see below). Febrile convulsions are common in hospitalized children (20–50%).[18]

Influenza in susceptible subgroups

Table 6 summarizes the high-risk populations for significant morbidity and mortality from influenza. These groups should be targeted for vaccination and other means of prophylaxis, as well as optimal disease management. Additional prophylaxis strategies include encouraging vaccination of close contacts of these individuals. These include family members and other household or institutional contacts.

People at high risk of hospitalization or death from influenza
People in specific age categories • adults aged 65 years and older • children under age 2 years
People with chronic disease • chronic pulmonary disease (asthma, bronchitis, emphysema, tuberculosis) • chronic cardiovascular disease • diabetes mellitus • renal disease
Immunocompromised people • transplant recipients • HIV-positive people • splenic dysfunction
Pregnant women

Table 6. People at high risk of hospitalization or death from influenza. Adapted from World Health Organization. Influenza vaccines. *Wkly Epidemiol Rec* 2000;**75**:281–288.

Children

Influenza attack rates are higher in children compared to adults. High-grade fever, cervical lymphadenopathy, and nausea and vomiting are frequent manifestations of influenza in children. Drowsiness (uncommon in adults) occurs in about 50% of infants under 4 years of age, and in up to 10% of 5–14-year-olds. Otitis media is frequently present (25% overall), particularly in those prone to middle-ear infections (67% in 1- to 3-year-olds). In infants (under 5 years old), 5–10 per thousand are hospitalized with respiratory symptoms (acute bronchitis, croup, pneumonia), and one in 5000 dies during influenza epidemics.

Infants and children under the age of 2 years will generally have a naive immune system and potentially compromised ability to mount an effective immune response to influenza. Morbidity and mortality are highest in this paediatric population. Additional common risk factors are related to asthma and type 1 diabetes, as well as immunocompromising conditions, although these are less common. School absentee rates can exceed 50% during influenza epidemics and provide evidence of the highly contagious nature of this illness. The diagnostic accuracy of influenza in children and the relationship to diagnostic testing has been recently reviewed.[19]

Older adults and adults with chronic diseases

Elderly patients experience more lower respiratory tract symptoms, with productive cough, wheeze and chest pain. In older adults, cough is the most common presenting symptom (>80% of cases), while fever is much less common compared to young adults. Considering that older people may lose 2–3% of muscle power per day of bed rest, significant disability may result from influenza. It is therefore not surprising that influenza and pneumonia represent one of the six leading causes of catastrophic disability in the population of 65 years and above.[20] Gastrointestinal symptoms also tend to be more common, especially with influenza B. Between four and 10 elderly patients per thousand are hospitalized. Recent studies have linked annual outbreaks of influenza to excess mortality during the winter months due to heart attacks, strokes, and exacerbations of diabetes and congestive heart failure during the winter months in the population of 70 years and older.[21,22] These observations point to a potentially enhanced inflammatory basis of influenza illness and related complications in older people.

As indicated above, influenza also poses increased risks to elderly (>65 years old) with chronic medical conditions. In these patients, influenza may occur as a serious illness, with exacerbation of the underlying disease and possibly a fatal outcome (reviewed in Ref. 23). Cough is the most common presenting symptom of influenza in this population and presents a diagnostic challenge to the physician.[24] More than 90% of influenza-associated excess deaths occur in the elderly of 65 years and older, and thus prevention, diagnosis and appropriate management of illness have a significant impact on health outcomes in this population.[25] People aged 50–65 years are also at greater risk due to the increased rate of chronic medical conditions, such as diabetes and cardiac, pulmonary or renal diseases, and immunocompromising conditions or medications. All of these conditions are associated with a significant increase in risk of hospitalization and death due to influenza, and also require optimal disease management along with vaccination for the prevention of influenza illness.

Immunocompromised patients

Influenza is more common in immunocompromised individuals (32%) than community controls (14%) and can cause severe disease.[26] Pneumonia (usually viral) is more common and mortality is higher. In a study of adults with leukaemia and adult recipients of haematopoietic stem cell transplants, severe lymphopenia, elderly age, receipt of an allogeneic haematopoietic stem cell transplant and concurrent opportunistic infections were identified as independent risk factors for a complicated course. Prolonged shedding of influenza virus (up to 5 months) and rapid emergence of resistance to amantadine and rimantadine (see Chapter 7) have been observed in immunocompromised individuals.[27] HIV infection is not generally associated with more severe influenza illness.

Pregnant women

Women in the second or third trimester of pregnancy have an increased risk of hospitalization due to cardiorespiratory complications.[28] The risk is greatest in pregnant women with mitral valve disease, who have a 45% mortality rate associated with influenza infection (higher if labour occurs during influenza infection).

Diagnosis of influenza

The accuracy of diagnosis of influenza tends to be strongly dependent on the physician's awareness of whether or not there is influenza activity in the local community. In clinical trials, accuracy of diagnosis during epidemics reached 62%, but may be as reliable as near-patient rapid testing (see below) when influenza is known to be circulating in the local community. Given the non-specific clinical presentation, the accuracy of diagnosis during epidemics is quite good. However, outside epidemic periods diagnostic accuracy may be less than 10%. Thus, it is important that primary-care physicians be aware of local influenza surveillance data (see list of websites in Appendix 1).

Criteria for clinical diagnosis of influenza	
Fever (≥ 37.8°C) and/or feverishness	
plus	
TWO of:	
• Cough	• Sore throat
• Nasal symptoms	• Myalgia
• Headache	• Malaise

Table 7. Criteria for clinical diagnosis of influenza. Reproduced from Nicholson KG. *Managing Influenza in Primary Care*, 1999 with permission from Blackwell Publishing.

The differential diagnosis of an influenza-like illness (ILI) includes a variety of viruses (e.g. respiratory syncytial virus, parainfluenza virus, rhino- and coronaviruses, adenoviruses) and bacteria (e.g. mycoplasma, chlamydia, rickettsia).

Proper and rapid diagnosis is essential to control influenza infection by antiviral treatment and to avoid the inappropriate use of antibiotics – the latter is a perennial problem, particularly in southern European countries. Accurate diagnosis is also very important for selection of vaccine virus strains and for the purposes of influenza surveillance.

Table 7 lists the criteria for purposes of clinical diagnosis of influenza that have been used in trials of antiviral drugs. Additional criteria are used in epidemiological studies to estimate attack rates and vaccine effectiveness, and to define clinical case definitions during the peak influenza period.[29] The most specific clinical case definition of influenza is "febrile upper respiratory illness" (URI), defined as 2 days of URI symptoms (runny nose, sore throat, cough) with two or more symptoms on at least 1 day and fever on at least 1 day.

A case definition of intermediate sensitivity and specificity is "any URI", defined as 2 days of at least one URI symptom (runny nose, sore throat, cough). The most sensitive clinical case definition of influenza is "any URI symptom", defined as any 1 day with at least one symptom (fever, runny nose, sore throat, cough, headache, muscle aches, chills, tiredness/weakness). The specificity of these case definitions is highly dependent on the attack rates of influenza in the community.

Thus, in conclusion, up-to-date tracking and awareness of circulating influenza in the local community is the single most important factor in the diagnostic accuracy of influenza as the cause of an acute respiratory tract illness.

Diagnostic tests

Diagnostic tests available for influenza include virus isolation and culture, antigen or viral RNA detection, and serology. However, the sensitivity and specificity vary widely, and the positive and negative predictive value of these tests is highly dependent on whether or not influenza is circulating in the local community – the contribution of these tests to clinical diagnosis thus needs to be carefully considered.[30] Rapid influenza tests have been developed but have some limitations (see below under "Antigen detection"). Appropriate samples for influenza testing include nasopharyngeal or throat swabs, nasal wash or nasal aspirates, depending on which type of test is used. Samples should be collected within the first 4 days of illness. Acute and convalescent serum samples can be tested for influenza antibodies. Viral culture or reverse transcriptase–polymerase chain reaction (RT-PCR) remains essential for determining the influenza subtypes and strains causing illness.

Serology

Influenza viruses cause agglutination of erythrocytes due to the capacity of the viral HA to bind to sialic acid residues on the red blood cell surface (see Chapter 2). Anti-HA antibodies interfere with this process and the haemagglutination–inhibition (HI) test is based on these properties (see Chapter 8). A four-fold or greater increase in HI antibodies is indicative of infection, occurring in about 70–90% of cases. Sera must be pre-treated to prevent false-positive results. The HI assay is used routinely to assess the response to vaccination.

Virus isolation and culture

The influenza virus can be propagated from clinical specimens on cultured cells or in embryonated chicken eggs. Because influenza may not produce cytopathic effects in cell culture, the haemagglutinating activity of the virus (see above) in the culture supernatant is used as a measure to screen for viral replication. Also in eggs, virus replication produces haemagglutinating activity. The entire process of virus culture and detection takes about 4–5 days, although more rapid (less sensitive) methods are available.

Antigen detection

Antigen can be detected by a variety of immunological techniques, including direct immunofluorescence of respiratory secretions, indirect antibody staining of exfoliated nasal epithelium or commercially available rapid influenza diagnostic tests. The latter differ in their ability to detect and distinguish between influenza A and B virus infections, methodologies, processing time and cost. The sensitivities of the rapid tests are lower than viral culture of respiratory specimens and may be comparable to symptom-defined illness when influenza is known to be circulating in the local community; a negative result might not exclude influenza virus infection. Rapid influenza tests provide results within 24 hours, while viral culture takes 4–5 days.

RNA detection by PCR

Although labour-intensive and usually slower than other tests, RT–PCR is used to type and subtype influenza infections.[31] Either clinical or cell culture specimens can be used. Multiplex PCR can differentiate between influenza A and B, parainfluenza 1, 2 and 3, and respiratory syncytial virus (RSV) with a very high degree of specificity (98%) and sensitivity (100%).

Influenza diagnosis in primary care

The diagnosis of influenza can be a particular challenge in the absence of surveillance data. Thus, the recommended approach for the patient who presents with an acute respiratory illness is to first determine whether or not a viral aetiology is likely according to the symptom criteria, and whether or not influenza is circulating in the local community. During an influenza outbreak, the primary-care physician can make the diagnosis of influenza based on clinical diagnostic criteria with similar diagnostic accuracy to that provided by the current rapid diagnostic tests. However, these tests may be helpful in the early stage of an influenza outbreak to identify sentinel cases. In addition, the collection of patient samples for virus culture and identification is critical to maintaining the network for influenza surveillance that determines the onset, peak and resolution of an influenza activity in the community.[32] The primary mode of influenza prevention is vaccination (see Chapter 8). Antiviral drugs (discussed in Chapter 7) provide additional prophylaxis, particularly in high-risk patients for whom influenza vaccination is contraindicated, and effective use of these drugs in the treatment of influenza depends on the physician's confidence in the diagnosis of influenza.[33]

References

1. Guan Y, Poon LL, Cheung CY *et al.* H5N1 influenza: a protean pandemic threat. *Proc Natl Acad Sci USA* 2004; **101**: 8156–8161.

2. Taubenberger JK. The virulence of the 1918 pandemic influenza virus: unraveling the enigma. *Arch Virol Suppl* 2005; **19**: 101–115.

3. Garcia-Sastre A. Antiviral response in pandemic influenza viruses. *Emerg Infect Dis* 2006; **12**: 44–47.

4. Nicholson KG, Wood JM, Zambon M. Influenza. *Lancet* 2003; **362**: 1733–1745.

5. Glezen WP, Taber LH, Frank AL, Gruber WC, Piedra PA. Influenza virus infections in infants. *Pediatr Infect Dis J* 1997; **16**: 1065–1068.

6. McBean AM, Hebert PL. New estimates of influenza-related pneumonia and influenza hospitalizations among the elderly. *Int J Infect Dis* 2004; **8**: 227–235.

7. Loeb M, McGeer A, McArthur M, Peeling RW, Petric M, Simor AE. Surveillance for outbreaks of respiratory tract infections in nursing homes. *Can Med Assoc J* 2000; **162**: 1133–1137.

8. Kaiser L, Fritz RS, Straus SE, Gubareva L, Hayden FG. Symptom pathogenesis during acute influenza: interleukin-6 and other cytokine responses. *J Med Virol* 2001; **64**: 262–268.

9. Vesa S, Kleemola M, Blomqvist S, Takala A, Kilpi T, Hovi T. Epidemiology of documented viral respiratory infections and acute otitis media in a cohort of children followed from two to twenty-four months of age. *Pediatr Infect Dis J* 2001; **20**: 574–581.

10. Reid AH, Taubenberger JK. The 1918 flu and other influenza pandemics: "over there" and back again. *Lab Invest* 1999; **79**: 95–101.

11. Webster RG. 1918 Spanish influenza: the secrets remain elusive. *Proc Natl Acad Sci USA* 1999; **96**: 1164–1166.

12. Cate TR. Clinical manifestations and consequences of influenza. *Am J Med* 1987; **82**: 15–19.

13. Ison MG, Hayden FG. Viral infections in immunocompromised patients: what's new with respiratory viruses? *Curr Opin Infect Dis* 2002; **15**: 355–367.

14. Berry L, Braude S. Influenza A infection with rhabdomyolysis and acute renal failure – a potentially fatal complication. *Postgrad Med J* 1991; **67**: 389–390.

15. Miura M, Asaumi Y, Wada Y *et al.* A case of influenza subtype A virus-induced fulminant myocarditis: an experience of percutaneous cardio-pulmonary support (PCPS) treatment and immunohistochemical analysis. *Tohoku J Exp Med* 2001; **195**: 11–19.

16. Kimura S, Ohtuki N, Nezu A, Tanaka M, Takeshita S. Clinical and radiological variability of influenza-related encephalopathy or encephalitis. *Acta Paediatr Jpn* 1998; **40**: 264–270.

17. Lasky T, Terracciano GJ, Magder L *et al.* The Guillain–Barré syndrome and the 1992–1993 and 1993–1994 influenza vaccines. *New Engl J Med* 1998; **339**: 1797–1802.

18. Heininger U. An update on the prevention of influenza in children and adolescents. *Eur J Pediatr* 2003; **162**: 828–836.

19. Uyeki TM. Influenza diagnosis and treatment in children: a review of studies on clinically useful tests and antiviral treatment for influenza. *Pediatr Infect Dis J* 2003; **22**: 164–177.

20. Ferrucci L, Guralnik JM, Pahor M, Corti MC, Havlik RJ. Hospital diagnoses, Medicare charges, and nursing home admissions in the year when older persons become severely disabled [see comments]. *J Am Med Assoc* 1997; **277**: 728–734.

21. Reichert TA, Simonsen L, Sharma A *et al.* Influenza and the winter increase in mortality in the United States, 1959–1999. *Am J Epidemiol* 2004; **160**: 492–502.

22. Fleming DM, Cross KW, Pannell RS. Influenza and its relationship to circulatory disorders. *Epidemiol Infect* 2005; **133**: 255–262.

23. Loeb M. Pneumonia in the elderly. *Curr Opin Infect Dis* 2004; **17**: 127–130.

24. Widdicombe J, Kamath S. Acute cough in the elderly: aetiology, diagnosis and therapy. *Drugs Aging* 2004; **21**: 243–258.

25. Janssens JP, Krause KH. Pneumonia in the very old. *Lancet Infect Dis* 2004; **4**: 112–124.

26. Vilchez RA, Fung J, Kusne S. The pathogenesis and management of influenza virus infection in organ transplant recipients. *Transpl Infect Dis* 2002; **4**: 177–182.

27. Englund JA, Champlin RE, Wyde PR *et al*. Common emergence of amantadine- and rimantadine-resistant influenza A viruses in symptomatic immunocompromised adults. *Clin Infect Dis* 1998; **26**: 1418–1424.

28. Lim WS, Macfarlane JT, Colthorpe CL. Treatment of community-acquired lower respiratory tract infections during pregnancy. *Am J Respir Med* 2003; **2**: 221–233.

29. Nichol KL, Mendelman P. Influence of clinical case definitions with differing levels of sensitivity and specificity on estimates of the relative and absolute health benefits of influenza vaccination among healthy working adults and implications for economic analyses. *Virus Res* 2004; **103**: 3–8.

30. File TM. Community-acquired pneumonia. *Lancet* 2003; **362**: 1991–2001.

31. Templeton KE, Scheltinga SA, Beersma MF, Kroes AC, Claas EC. Rapid and sensitive method using multiplex real-time PCR for diagnosis of infections by influenza A and influenza B viruses, respiratory syncytial virus, and parainfluenza viruses 1, 2, 3, and 4. *J Clin Microbiol* 2004; **42**: 1564–1569.

32. Meerhoff TJ, Meijer A, Paget WJ. Methods for sentinel virological surveillance of influenza in Europe – an 18-country survey. *Eur Surveill* 2004; **9**: 1–4.

33. Potter CW. Chronicle of influenza pandemics. In: Nicholson KG, Webster RG, Hay AJ, editors. *Textbook of Influenza*. Oxford: Blackwell Science, 1998; pp. 3–18.

6. Social and Economic Impact of Influenza

The societal burden of recurring annual influenza epidemics and, in particular, occasional pandemics is often underestimated. It is important to understand the extent to which influenza impacts on society, its financial burden and strategies to minimize the impact. This chapter provides an overview of the social and economic implications of influenza. It introduces economic concepts and methods, and examines the impact of influenza on different population groups (including children, working adults and the elderly) and sectors of the economy (health care and work place).

Each outbreak of influenza takes a significant toll on societies in terms of morbidity, mortality and financial resources.[1,2] For example, the WHO estimates that there are 3–5 million cases of severe influenza illness, resulting in 250,000–500,000 deaths annually in the industrialized world. Vaccination is the most important strategy for preventing influenza. Yet, data on vaccination coverage as discussed in Chapter 8, show that even in industrialized countries, large proportions of the population at risk do not receive the influenza vaccine. The target of the WHO of reaching 75% vaccination coverage of high-risk people (see Chapter 8) therefore remains an important goal for 2010. Achieving this target will also have major positive implications for the capacity of individual countries to respond adequately to influenza pandemics. The chapter finishes with a brief look at the social and economic impact of influenza pandemics. The health benefits as well as the cost-effectiveness of influenza vaccination will be discussed in Chapter 8.

Social impact

In addition to the short-term incapacitating symptoms of influenza, complications may arise from having influenza either directly and indirectly (see Chapter 5). These

Key Messages

- The WHO estimates that 3–5 million cases of severe influenza illness result in 250,000–500,000 deaths annually in the industrialized world.
- Compared to healthy adults, influenza-related death rates are as much as 50- to 100-fold higher among patients with cardiovascular or pulmonary disease, and even higher for those with two or more co-morbid conditions.
- In young children, influenza is associated with increased rates of otitis media and serious cardiorespiratory illnesses leading to hospitalization.
- The annual cost of influenza is highly variable and may be as high as €20 million in direct health-care costs in some European countries – these direct costs are less than 10% of the total annual cost to society.
- Work absenteeism from influenza illness results in indirect costs that may also be as high as €20 million in lost days of work in a single influenza season.
- School absenteeism not only impacts on children but contributes on average to a loss of 3 work days for the parent who must remain at home with the child.
- Intangible costs include those related to death, increased disability or a diminished quality of life for the individual and the potential additional burden to caregivers.
- The Centers for Disease Control (CDC) has estimated that in the USA, during an influenza pandemic, high-risk individuals representing 15% of the population would account for ~84% of all deaths, at a cost of US $71.3–166.5 billion, excluding disruptions to commerce and society.

complications have a large social impact, including increased demands on health-care systems, long-term illness and disability, or even mortality.

Epidemiological records on the morbidity and mortality statistics related to influenza show that, in the winters from 1990 to 1998, the mean annual estimate of influenza-associated deaths in the USA was 51,203, mostly associated with influenza A (H3N2).[3] The death rates varied from year to year, with an average across the decade of 195 influenza-related deaths per million population (Table 8).

The influenza attack rate varies from year to year as does the virus strain. The annual attack rate lies between 6–12% of the population. In the USA, in recent years, an annual average of approximately 36,000 people die and 200,000 people are hospitalized due to influenza.[4] Based on a total population of 298 million, these figures represent mortality and hospitalization rates of 120 and 670 per million population, respectively. The current mortality rate is thus lower than the average excess mortality in the USA during the last decade (Table 8). Extrapolating the current numbers for the USA, Table 9 shows the average annual excess influenza-associated deaths and hospitalizations for a number of countries in Europe.

Estimated annual influenza deaths rates per 100,000 in the USA (1990–1999)

Season	A/H1N1	A/H3N2	B	Total
1990–1991	0.79	2.41	7.00	10.20
1991–1992	2.57	18.11	0.22	20.90
1992–1993	0.46	7.75	7.42	15.64
1993–1994	0.07	18.88	0.16	19.10
1994–1995	0.22	12.91	2.72	15.85
1995–1996	5.58	8.94	2.84	17.36
1996–1997	0.00	20.98	4.73	25.71
1997–1998	0.02	26.27	0.24	26.54
1998–1999	0.11	20.39	3.57	24.07
Mean	**1.09**	**15.18**	**3.21**	**19.49**

Table 8. Estimated annual influenza deaths rates per 100,000 in the USA (1990–1999). Reproduced from Thompson WW *et al.* Mortality associated with influenza and respiratory syncytial virus in the United States. *JAMA*; **289**: 179–186[3] with permission from the American Medical Association.

Estimated averages of annual excess morbidity and mortality in Europe during influenza epidemics			
Country	**Inhabitants (millions)**	**Excess deaths**	**Excess hospitalizations**
France	60.5	7260	40,535
Germany	82.7	9924	55,409
Netherlands	16.3	1956	10,921
Spain	43.1	5172	28,877
Greece	11.1	1332	7437
Italy	58.1	6972	38,927
UK	59.7	7164	39,999
Sweden	9.0	1080	6030
Ukraine	46.5	5580	31,155
Poland	38.5	4620	25,795

Table 9. Estimated averages of annual excess morbidity and mortality in Europe during influenza epidemics. The numbers are extrapolated from the current USA estimates of 120 excess deaths and 670 excess hospitalizations per million population. Centers for Disease Control and Prevention (CDC). Key facts about influenza and the influenza vaccine. (http://www.cdc.gov/flu/keyfacts.htm)[4]

Population groups at higher risk

As discussed in the previous chapter, complications arising from influenza are more frequent and serious among adults with chronic conditions, such as pneumonia or respiratory disease, cardiac disease or cerebrovascular disease.[5] Influenza may cause secondary bacterial pneumonia and exacerbations of underlying chronic medical conditions that can result in hospitalization or even death. These groups are at a higher morbidity and mortality risk from influenza.

Death rates associated with influenza and pneumonia have been estimated as two per 100,000 for patients with no high-risk conditions, 10 per 100,000 in those with one high-risk condition and 377 per 100,000 in those with two or more high-risk conditions.[6] Other studies have reported that influenza-related death rates are as high as 50 times greater in patients with cardiovascular disease, 100 times greater in patients with pulmonary disease, and even higher in those with two or more co-morbid conditions, than in healthy adults.[7]

Age poses an additional risk factor. Even in the absence of underlying medical conditions, the elderly are at increased risk for serious complications and possibly death due to influenza, primarily as a result of a decline in cell-mediated immunity with age, as discussed in Chapters 4 and 5. Furthermore, the incidence of chronic medical conditions increases with age, adding to the risk. It is for these reasons that fatalities due to influenza rise with age and are highest among elderly people with chronic medical conditions – indeed, in this group up to 90% of influenza-related deaths are observed.

Children are also at increased risk for complications due to influenza, in particular small infants because of their immature immune system (see Chapter 5). As with the elderly, a chronic underlying medical condition adds to the risk. Thus, among children, the largest high-risk group involves those with asthma, accounting for approximately 7% of all children. These children have considerably more outpatient visits, antibiotic courses and higher hospitalization rates during all seasons than other children at increased risk.[8] In young children, influenza may be associated with increased rates of otitis media and, even more serious, cardiorespiratory illnesses leading to hospitalization (Table 10).

Economic impact

In addition to its social impact, influenza also has a significant economic impact, including lost or reduced productivity in the work place or at home due to incapacitating illness, and increased health-care costs from treating the disease. Accordingly, the economics of

Influenza-related health-care interventions among children	
	Excess number per 1000 children
Hospitalizations for children:	
• <1 year of age	19
• 1 to ≤ 3 years of age	8
• Outpatient visits	200
• Courses of antibiotic therapy	140

Table 10. Influenza-related health-care interventions among children. Reproduced from Neuzil KM *et al*. The burden of influenza illness in children with asthma and other chronic medical conditions. *J Pediatr* 2000; **137**: 856–864[8] with permission from Elsevier.

influenza are concerned primarily with two issues: the economic impact of influenza and the cost of interventions to prevent or treat influenza.

Understanding the economics of influenza

The economic impact of disease is often measured by conducting a cost-of-illness study, which estimates the overall cost of the disease, including its treatment and consequences. These costs can be classified into three separate types:

- Direct costs – these are actually incurred monetary costs, and imply a cash outlay (acquisition costs, fees, charges).
- Indirect costs – these are not paid in cash but can be inferred, as they arise from reduced or lost productivity due to work absenteeism, decreased performance or even premature death.
- Intangible costs – these are attributable to pain, suffering and reduced quality of life. Avoidance of these is often identified as an intangible benefit. These costs cannot be easily converted to a monetary value.

In influenza, the cost of prevention or treatment strategies, such as influenza vaccination programmes, may be measured using cost–effectiveness analyses or cost–benefit analyses. These analyses of influenza-related health-care interventions measure whether the benefits derived from the intervention are worth the cost of undertaking the intervention.

The costs of the intervention are typically measured from the perspective of the payer (e.g. health-care system) and include the direct cost of the intervention itself. The benefits, on the other hand, can include direct, indirect and intangible costs resulting from the intervention. These may include reduced hospitalizations and physician visits, reduced work or school absenteeism, reduced mortality and improved quality of life. Reduced health-care visits can provide a cost saving to the health-care payer. Reduced absenteeism at work is an indirect benefit to the employer and society. Reduced mortality and improved quality of life is an intangible benefit that is gained by the individual, their family and friends, and society at large.

Economic evaluations estimating the cost-effectiveness of programmes should always compare one health-care strategy against another. For example, when estimating the cost-effectiveness of influenza vaccination, the analysis may compare the expected overall cost and benefits of no vaccination against the expected costs and benefits resulting from an influenza vaccination programme. The decision-maker would then be presented with the additional costs of the vaccination programme and determine if it is worth the additional cost given the additional benefits it provides during the influenza season.

The results from such analyses are usually presented in the form of an incremental cost-effectiveness ratio (ICER), which shows the additional cost of providing the intervention per extra additional unit of benefit. This can be illustrated with a simple hypothetical example. Suppose a programme to vaccinate 1000 people costs €100,000. The vaccination reduces health-care resource use, saving the health-care payer €50,000. Furthermore, by reducing the number of people with influenza, the programme also avoids five deaths relating to influenza. Therefore, the ICER would be:

(€100,000 – €50,000)/5 = €10,000 per life saved.

If the decision-maker would normally be prepared to pay more than €10,000 to save a life, the programme would be deemed a cost-effective option. If the programme can avoid more costs than it costs to implement, and it saves more lives, then the programme is always cost-effective as it provides more benefit and saves money.

A cost–benefit analysis converts all effectiveness benefits, including intangible costs, into a monetary value. In the example above, the decision-maker may value a life saved at €30,000. Consequently, the monetary benefit of the programme is €50,000 (health-care savings) plus €150,000 (five lives saved at a value of €30,000 each). This total monetary benefit of €200,000 outweighs the €100,000 cost of implementing the programme. The programme would therefore be deemed cost-effective and should be implemented.

Cost of illness

The increasing cost of health care has prompted research to quantify the cost of illness due to influenza. Cost-of-illness research is the first step in resource allocation decision-making. It results in an increased awareness of the impact of infection, it determines the burden of disease in relation to other diseases, it defines the unmet medical need which future drug development should address, and it facilitates the formulation of prevention, vaccination and treatment policies. Cost-of-illness research is hampered by underestimates of the incidence of the disease due to influenza and difficulties quantifying indirect and intangible costs. A study in France has shown that the cost of influenza was up to US $20 billion during the 1988/89 epidemic (Table 11).[9,10]

Direct costs

The direct costs of influenza have been estimated in a number of different studies. These vary depending on the structure of the national health services. Physician visits act as the main cost driver. In the USA, influenza affects approximately 50–60 million people annually, of whom about 50% visit their doctor.[11] Most of these 25–30 million physician visits are by schoolchildren and young adults. Between 114,000 and 142,000 (2.1–2.6 per 1000) of those infected are hospitalized, mainly very young children (5 per 1000) and elderly people (3–4 per 1000).[12] The annual direct medical costs of influenza in the USA are estimated at between $3 and $5 billion.[13]

Indirect costs: work productivity loss and school absenteeism

During influenza outbreaks, there is a significant rise in absenteeism in the working population. Indeed, the most significant cost of influenza to society is the indirect cost related to lost productivity and work absenteeism. Reported studies of work loss due to influenza range from 0.79 to 4.9 days of work lost.[14] From a UK study, the mean time off work for health-care workers with influenza was 4 days.[15] However, it has been also been noted that health-care workers tend to continue to work during their illness.[16] From the National Health Interview Survey in the USA, influenza was responsible for more than 200 million days of restricted activity, 100 million days of bed disability, 75 million days of lost work and 22 million health-care provider visits across the USA in 1995.[14] Surveillance programmes have found that the high absenteeism in the work place is common during peak influenza activity, typically between December and March.[17] Overall, influenza accounts for around 10% of sickness absence from work.

Estimates of the cost of influenza have shown that indirect costs can be as much as 10 times higher than the direct medical costs and range from $10 to $15 billion per year.[9,10] This is illustrated in Table 11, which presents the estimated direct and indirect costs of the moderate 1989

Estimated cost of an influenza epidemic in France in 1989

Costs	1989 (French francs, millions)		2005 equivalent (€, millions)	
	National health insurance system	Society	National health insurance system	Society
Direct medical costs				
• GP visits	363.95	494.20	423.93	575.65
• Medication	350.69	637.61	408.49	742.70
• Hospitalizations	92.39	115.49	107.62	134.52
Total direct medical costs	807.03	1247.31	940.04	1452.87
Indirect costs				
Cost of absenteeism and loss of productivity	1121.30	13,076.00	1306.10	15,231.11
Total costs	**1928.33**	**14,323.31**	**2246.14**	**16,683.98**

Table 11. Estimated cost of an influenza epidemic in France in 1989. Adapted from Levy E. French economic evaluations of influenza and influenza vaccination. *Pharmacoeconomics* 1996; **9**(Suppl 3): 62–66[9] with permission from Adis International.

influenza epidemic in France. In this example, the indirect costs due to work absenteeism and loss of productivity from the societal perspective, account for the major part (91%) of the estimated cost of illness. Obviously, from the perspective of the French national health insurance (Table 11), the indirect costs related to absenteeism and loss of productivity represent a much smaller share (58%).

Numerous studies have been conducted around the world estimating the indirect cost due to influenza-like illness (ILI) in the work place. In the general population, it takes 7 days on average to alleviate the major symptoms of ILI.[18] From a review of the literature, Nichol concluded that workers with ILI took an average of 2 days off work.[14] This was based on estimates in the literature ranging from 0.79 days[19] to 4.9 days.[20] Nichol *et al.*[21] estimated that for every 100 non-vaccinated employees, 52 working days were lost due to ILI. Campbell and Rumley[22] estimated the direct costs of medical care received for ILI at $22.80 per person, and the indirect costs at $80 to the company for each day's absence from work. Other estimates for the cost per working day lost are higher; for example, in Dille,[17] the cost per work day lost was estimated at $175.24.

Even a mild attack of influenza can reduce reaction times by 20–40%,[23] with associated health and safety implications among those who continue working despite influenza infection (up to half of employees). Workers who have ILI and either stay at work or return to work before the illness has disappeared have reduced effectiveness (lower productivity) while working. Keech *et al.*[24] calculated that workers, on average, returned to work 0.7 days before they should. As well as reduced effectiveness at work, there are additional burdens associated with making poor decisions when one is ill.

Burckel *et al.*[25] calculated mean attack rates for different levels of morbidity for various age groups based on information from Houston Surveillance and European evidence, assuming an overall attack rate of 10%. The 18–24 age group was found to be most likely to contract low-morbidity influenza in which normal work schedules are essentially maintained and the related impairment was equivalent to an average of one lost day of work. In moderate influenza morbidity, 3 days of work were lost for patients under 45 years and 4 days for those over 45 years. High-morbidity influenza was defined as influenza cases that require hospitalization with an average of 12 days of work lost.[25]

In a study of families with school-aged children during an influenza season compared with a non-influenza winter season, there was a significantly greater total number of illness episodes, febrile illnesses, analgesic use, school absences, parental absenteeism and secondary illnesses among family members. It was concluded that the influenza season has significant adverse effects on the quality of life of school-aged children and their families.[26] Time off work due to children being away from school with influenza can also be an important factor. On average, 3 days of work are lost due to a child being absent from school because of influenza, requiring a parent to remain at home.[14]

Intangible costs

Intangible costs include quality of life and impairment of function and performance. Influenza infection is stressful for the individual infected and frequently places a burden of care on other family members. This can restrict the usual activities of families and reduce the quality of life of the affected individual, as well as those around him/her. The socio-economic impact of these factors is difficult to quantify. In elderly people, in particular, infection is associated with a decline in major physical functions.[27]

Impact of pandemic influenza

While the costs of epidemic influenza are considerable, the potential social and economic impact of pandemic influenza is even greater. It is difficult to predict the impact of a future pandemic with any reasonable degree of accuracy. Factors influencing the impact include the attack rate, the pathogenicity of the virus and the adequacy of control measures. Pandemic influenza results from the emergence and rapid spread of a novel influenza virus, as discussed in Chapter 3. The most devastating on record was the 1918–19 "Spanish flu" pandemic caused by an H1N1 virus, estimated to have caused 50 million deaths worldwide (see Chapter 3). While the 1957 Asian flu (H2N2) and the 1968 Hong Kong flu (H3N2) have caused less severe pandemics, the morbidity and mortality among the elderly and people with high-risk conditions remained severe.

The US Center for Disease Control and Prevention (CDC) has estimated the costs involved in the event of a pandemic in the USA.[28] Patients at high risk (15% of the population) would account for approximately 84% of all deaths, and the economic impact would be between US $71.3 and $166.5 billion, excluding disruptions to commerce and society (Table 12).

In the event of another pandemic, and particularly in light of rapid ageing of the population, a crisis in health-care capacity and ability to cope with the emergency may be anticipated along with a considerable social and

Estimated socio-economic burden of a future influenza pandemic in the USA	
Clinical event	**Numbers affected**
Deaths	89,000–207,000
Hospitalizations	314,000–734,000
Outpatient visits	18–42 million
Additional illnesses	20–47 million

Table 12. Estimated socio-economic burden of a future influenza pandemic in the USA. Data derived from Meltzer MI *et al*. The economic impact of pandemic influenza in the United States: priorities for intervention. *Emerg Infect Dis* 1999; **5**: 657–671.[28] Available from http://www.cdc.gov/ncidod/eid/vol5no5/meltzer.htm

economic impact.[29] One study using a 25% attack rate in the event of a pandemic estimated a 3.7-fold rise in outpatient visits, a 3.9-fold increase in hospitalizations and an 8.2-fold rise in deaths.[30] Clearly, this would outstrip the capacity of the current supply system and place an enormous burden on the supply management for vaccine, as discussed further in Chapter 9.[31]

References

1. Sullivan KM. Health impact of influenza in the United States. *Pharmacoeconomics* 1996; **9**(Suppl 3): 26–33.

2. Szucs T. The socio-economic burden of influenza. *J Antimicrob Chemother* 1999; **44**: 11–15.

3. Thompson WW, Shay DK, Weintraub E *et al.* Mortality associated with influenza and respiratory syncytial virus in the United States. *J Am Med Assoc* 2003; **289**: 179–186.

4. Centers for Disease Control and Prevention (CDC). Key Facts about Influenza and Influenza Vaccine. (www.cdc.gov/flu/keyfacts.htm)

5. Nichol KL, Wuorenma J, von Sternberg T. Benefits of influenza vaccination for low-, intermediate-, and high-risk senior citizens. *Arch Intern Med* 1998; **158**: 1769–1776.

6. Zimmerman R. Lowering the age for routine influenza vaccination to 50 years. *Am Fam Physician* 1999; **60**: 2061–2066.

7. Barker WH, Mullooly JP. A study of excess mortality during influenza epidemics in the United States, 1968–1976. *Am J Epidemiol* 1982; **115**: 479–480.

8. Neuzil KM, Wright PF, Mitchel EF, Griffin MR. The burden of influenza illness in children with asthma and other chronic medical conditions. *J Pediatr* 2000; **137**: 856–864.

9. Levy E. French economic evaluations of influenza and influenza vaccination. *Pharmacoeconomics* 1996; **9**(Suppl 3): 62–66.

10. Szucs TD. Influenza. The role of burden-of-illness research. *Pharmacoeconomics* 1999; **16**(Suppl 1): 27–32.

11. Couch RB. Influenza: prospects for control. *Ann Intern Med* 2000; **133**: 992–998.

12. CDC. Prevention and control of influenza. Recommendations of the Advisory Committee on Immunization Practices (ACIP). *MMWR Recomm Rep* 2001; **50**(RR-4): 1–46.

13. Patriarca PA. New options for prevention and control of influenza. *J Am Med Assoc* 1999; **282**: 75–77.

14. Nichol KL. Cost-benefit analysis of a strategy to vaccinate healthy working adults against influenza. *Arch Intern Med* 2001; **161**: 749–759.

15. Elder A, O'Donnell B, McCruden E *et al.* Incidence and recall of influenza in a cohort of Glasgow health care workers during the 1993–4 epidemic: results of serum testing and questionnaire. *Br Med J* 1996; **313**: 1241–1242.

16. Wilde J, MacMillan J, Serwint J *et al.* Effectiveness of influenza vaccine in health care professionals: a randomised trial. *J Am Med Assoc* 1999; **281**: 908–913.

17. Dille JH. A worksite influenza immunization program. Impact on lost work days, health care utilization, and health care spending. *AAOHN J* 1999; **47**: 301–309.

18. Mauskopf JA, Cates SC, Griffin AD. A pharmacoeconomic model for the treatment of influenza. *Pharmacoeconomics* 1999; **16**(Suppl 1): 73–84.

19. Bridges C, Thompson W, Melzer M *et al.* Effectiveness and cost-benefit of influenza vaccination of healthy working adults. *J Am Med Assoc* 2000; **284**: 1655–1663.

20. Kumpulainen V, Makela M. Influenza vaccination among healthy employees: a cost-benefit analysis. *Scand J Infect Dis* 1997; **29**: 181–185.

21. Nichol KL *et al.* The effectiveness of vaccination against influenza in healthy, working adults. *New Engl J Med* 1995; **333**: 889–893.

22. Campbell DS, Rumley MH. Cost-effectiveness of the influenza vaccine in a healthy, working-age population. *J Occup Environ Med* 1997; **39**: 408–414.

23. Monto AS, Sullivan KM. Acute respiratory illness in the community. Frequency of illness and the agents involved. *Epidemiol Inf* 1993; **110**: 145–160.

24. Keech M, Scott A, Ryan P. The impact of influenza and influenza-like illness on productivity and health care resource utilisation in a working population. *Occup Med* 1998; **48**: 85–90.

25. Burckel E, Ashraf T, Galvao de Sousa Filho J *et al.* Economic impact of providing workplace influenza vaccination. A model and case study application at a Brazilian pharma-chemical company. *Pharmacoeconomics* 1999; **16**: 563–576.

26. Neuzil KM *et al.* Illness among schoolchildren during influenza season: effect on school absenteeism, parental absenteeism from work, and secondary illness in families. *Arch Pediatr Adolesc Med* 2002; **156**: 986–991.

27. Barker WH, Borisute H, Cox C. A study of the impact of influenza on the functional status of frail older people. *Arch Intern Med* 1998; **158**: 645–650.

28. Meltzer MI, Cox NJ, Fukuda K. The economic impact of pandemic influenza in the United States: priorities for intervention. Centers for Disease Control and Prevention, Atlanta, Georgia, USA. *Emerg Infect Dis* 1999; **5**: 657–671.

29. Cox NJ, Tamblyn SE, Tam T. Influenza pandemic planning. *Vaccine* 2003; **21**: 1801–1803.

30. Schopflocher DP, Russell ML, Svenson LW, Nguyen TH, Mazurenko I. Pandemic influenza planning: using the U.S. Centers for Disease Control FluAid Software for small area estimation in the Canadian context. *Ann Epidemiol* 2004; **14**: 73–76.

31. Medema JK, Zoellner YF, Ryan J, Palache AM. Modeling pandemic preparedness scenarios: health economic implications of enhanced pandemic vaccine supply. *Virus Res* 2004; **103**: 9–15.

7. Antivirals: Treatment, Prophylaxis and Pandemic Control

Influenza often presents to the primary-care physician several days into an illness that begins with fever, myalgia and cough. Patients, on their own or at the recommendation of their physician, may use paracetamol, aspirin or other non-steroidal anti-inflammatory medications for symptomatic relief. The symptoms of influenza might be regarded by the patient as a bacterial infection and the reason for a visit to the physician is often to obtain a prescription for antibiotics. This and/or inaccurate diagnosis of the disease (see Chapter 5), may lead to over-prescribing of antibiotics for uncomplicated influenza illness, which contributes to the development of microbial resistance to antibiotics. Bacterial complications of influenza generally present 5–7 days after the onset of illness, during a phase of apparent recovery, with subsequent recurrence of symptoms (see Chapter 5).

Clearly, vaccination remains the method of choice for influenza prophylaxis (see Chapter 8). However, under specific conditions – for example, when vaccination has not or cannot be performed, perhaps because of hypersensitivity to egg proteins, or when vaccinated subjects are not fully protected – the use of antiviral drugs should be considered for treatment or prevention of influenza. Antiviral drugs effective against influenza include the M2 channel inhibitors amantadine (Symmetrel®, Lysovir®, Symadine®) and rimantadine (Flumadine®), and the neuraminidase inhibitors (NAIs) zanamivir (Relenza®) and oseltamivir (Tamiflu®). These drugs interfere with specific steps in the replication cycle of influenza virus, either at the level of virus entry or at the level of virus assembly or release from the infected cell.

Since M2 is not present in influenza B or C viruses, the M2 channel inhibitors are active only against influenza A. Their use has been limited by the occurrence of adverse side effects and the risk of induction of drug-resistant viral strains. The NAIs represent a novel class of anti-influenza drugs that are active against influenza A and B viruses; adverse effects are very limited and, thus far, significant development of resistant viral variants has not been reported for these drugs.

Mechanisms of action of antivirals

Amantadine and rimantadine were the first generation of influenza antiviral agents.[1] These compounds specifically

Key Messages
• Antibiotics are often inappropriately prescribed when uncomplicated influenza infection presents to the physician.
• Antiviral drugs must be initiated within 24–48 hours of the onset of symptoms for the treatment of influenza.
• Antiviral drugs must be taken daily through the period of exposure risk for effective prophylaxis against influenza.
• The use of M2 inhibitors has been limited by the induction of drug-resistant strains of influenza.
• Particularly in older people, the use of M2 inhibitors has been limited by their adverse effects and need for dose adjustment due to renal excretion of the drugs.
• In the case of neuraminidase inhibitors, there are thus far few observations of induction of drug-resistant virus strains, although the emergence of Tamiflu-resistant H5N1 virus has been reported.
• In contrast to the M2 inhibitors, neuraminidase inhibitors have fewer adverse side effects.
• In the initial phase of an influenza pandemic, antivirals will be the only means to combat its impact.
• Multiple drug strategies for prophylaxis and treatment may be needed to address the issues of antiviral resistance in an influenza pandemic.

block the ion channel function of the M2 protein of influenza A virus (see Chapter 2), thus interfering with corresponding specific steps in the viral life cycle. The neuraminidase inhibitors are novel drugs, designed on the basis of the three-dimensional structure of the influenza A and B neuraminidase[2] (see Chapter 2). The mechanisms of action of the four available specific anti-influenza viral drugs are summarized in Figure 22.

The M2 channel inhibitors – amantadine and rimantadine

At high concentrations (>15 μg/ml), amantadine and rimantadine non-specifically raise the pH within cellular endosomes, thus inhibiting or retarding the acid-induced conformational change in the viral HA. However, the required concentrations of the drugs are not generally attained *in vivo*. At low, pharmacologically relevant concentrations (<0.75 μg/ml), amantadine and rimantadine specifically inhibit the ion channel activity of the M2 protein, probably through direct binding to the pore region of the protein.[3] In doing so, the drugs inhibit acidification of the interior of susceptible viruses and dissociation of the M1 protein from the viral nucleocapsid (Figure 22), which is a necessary step in the uncoating of the viral genome during infection (see Chapter 2). In addition, within the infected cell, M2 protects newly synthesized HA against premature exposure to low pH by transiently neutralizing the pH of the trans-Golgi network, while HA is in transit to the cell surface.[4] Amantadine and rimantadine counteract this protection and thus indirectly induce a premature conformational change in HA, inactivating the protein (Figure 22).

The neuraminidase inhibitors – zanamivir and oseltamivir

Influenza A and B viruses (but not influenza C viruses) possess neuraminidase on their outer surface, an enzyme essential for release of virus particles from infected cells, for prevention of formation of viral aggregates and for viral spread within the respiratory tract.[2] Neuraminidase cleaves the receptor for influenza A and B viruses, sialic acid, from glycoproteins and glycolipids. Zanamivir and oseltamivir are analogues of sialic acid, as shown in Figure 23.[2] These

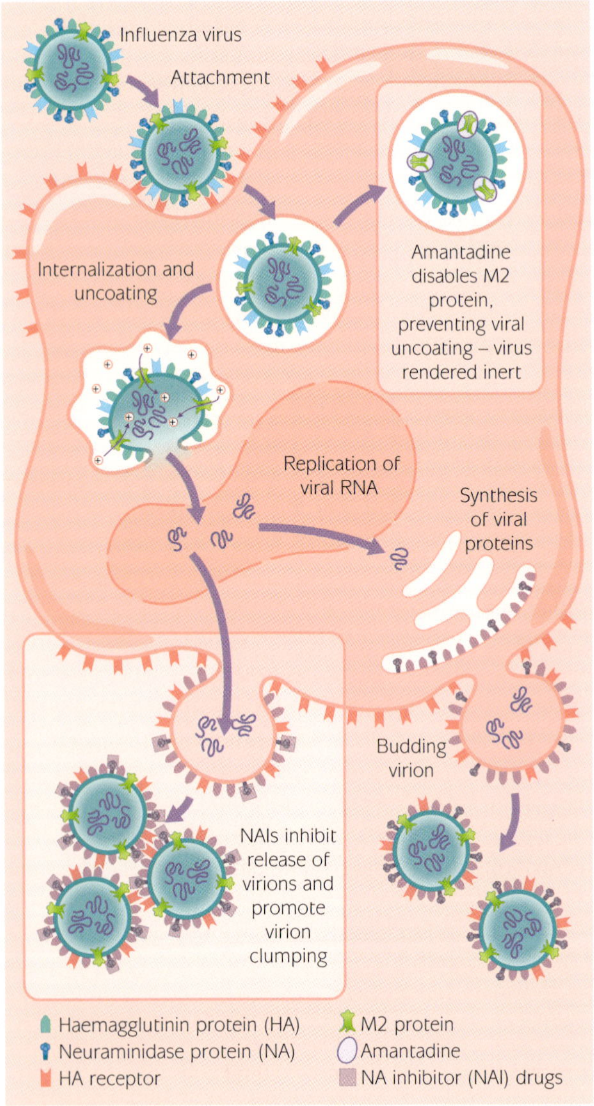

compounds specifically inhibit all nine NA subtypes in nature, including the subtypes contained in the avian strains of influenza A H5N1, H7N7 and H9N2 that have infected humans (see also Chapter 3).

The inhibition of neuraminidase on both influenza A and B viruses has three important consequences.[2] First, it hinders the passage of the virus through the mucus of the respiratory tract and thus retards initial infection of epithelial cells. Second, it inhibits the release of new viral particles from the surface of infected cells and thus slows the spreading of the virus through the respiratory tract of the host. Third, as inhibition of neuraminidase interrupts the release of virus particles from infected cells, it also inhibits transmission of the virus from one infected individual to another.

Antivirals in treatment and prophylaxis of influenza

Table 13 presents a comparison between the different antiviral drugs available for treatment (amantadine, rimantadine and oseltamivir) and prophylaxis (amantadine, rimantadine, oseltamivir and zanamivir) of influenza. Again, only oseltamivir and zanamivir are effective against both influenza A and B. The UK National Institute for Clinical Excellence (NICE) no longer approves the use of amantadine for treatment of influenza,[5] while the USA and Canada have also recently changed the recommendations for prophylaxis

Figure 22. The mechanism by which antiviral drugs interrupt the replicative cycle of influenza is illustrated. M2 inhibitors prevent the M2-mediated acidification of the interior of the virus while it resides in endosomes and the subsequent uncoating of the viral genome, thus inhibiting viral replication. Neuraminidase inhibitors (NAIs) prevent cleavage of sialic acid residues and thus newly formed virus cannot be released from the cell surface to infect adjacent cells; also, virus particles remain associated to one another. "The treatment of influenza with antiviral drugs" – Reprinted from *CMAJ* 07-Jan-03; **168(1)**: 49–57 by permission of the publisher. © 2003 CMA Media Inc.

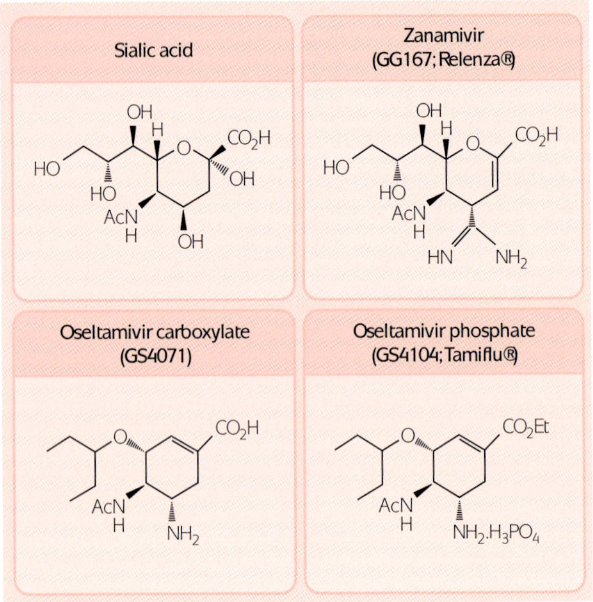

Figure 23. The neuraminidase inhibitors zanamivir and oseltamivir are structural analogs of sialic acid, which is the substrate of neuraminidase (NA) and the receptor for the influenza virus HA. Zanamivir and oseltamivir bind to the substrate binding site of NA, thus blocking its enzymatic activity. Zanamivir (GG167 or Relenza®) is inhaled or administered intranasally. Oseltamivir is given as the oral prodrug oseltamivir phosphate (GS4104 or Tamiflu®), which is converted in the liver to oseltamivir carboxylate GS4071, the active drug.

and treatment with amantidine and rimantidine, the reasons for which will be discussed below.

The influenza antiviral drugs should only be used to treat patients if the clinical picture meets the criteria for influenza-like illness (ILI) and if influenza activity has been reported in the area. Table 14 lists indications for influenza antivirals.

Table 13. Comparison of antiviral drugs for prophylaxis and treatment of influenza. Reproduced from La Rosa AM and Whimbey E. Respiratory viruses. In: Cohen J, Powderly WG, editors. *Infectious Diseases*, 2nd edn, 2003 with permission from Elsevier.

Comparison of antiviral drugs for prophylaxis and treatment of influenza

Drug	Trade name	Influenza type	Dosing for prophylaxis	Dosing for treatment	Main side effects
Amantadine	Symmetrel®	A	Age 1–9 years: 5 mg/kg/day, p.o. div b.i.d. Age 9 and up: 100 mg p.o. b.i.d.	Age 1–9 years: 5 mg/kg/day, p.o. div b.i.d. Age 9 and up: 100 mg p.o. b.i.d.	Central nervous system
Rimantadine	Flumadine®	A	Age 1–10 years: 5 mg/kg/day, p.o. q.d. Age 10 and up: 100 mg p.o. b.i.d.	Adults: 100 mg p.o. b.i.d.	Central nervous system
Zanamivir	Relenza®	A and B	N/A	Age >7 years: 10 mg inhaled b.i.d.	Bronchial
Oseltamivir	Tamiflu®	A and B	Age 1–12 years: dose per weight Age ≥ 13 years: 75 mg p.o. q.d.	Age 1–12 years: dose per weight Age 13 and up: 75 mg p.o. b.i.d.	Gastrointestinal

*For children who weigh <15 kg the dose is 30 mg b.i.d.; for those who weigh 15–23 kg, it is 45 mg b.i.d.; for those who weigh 23–40 kg, it is 60 mg b.i.d.; and for those who weigh >40 kg, it is 75 mg b.i.d.

Treatment of influenza

All influenza antiviral agents have to be administered within 24–48 hours of disease onset. The drugs are given orally, except for zanamivir, which is inhaled. Influenza antiviral drugs diminish the severity of clinical symptoms and shorten the duration of uncomplicated influenza disease by an average of 1 day (from about 7 to 6 days). To reduce the risk of development of drug resistance (see below), amantadine or rimantadine should be discontinued as soon as clinically appropriate, usually after 3–5 days or within 24–48 hours after the disappearance of symptoms. For zanamivir and oseltamivir, the recommended duration of treatment is 5 days.

Evidence for the effectiveness of these four antiviral agents is based primarily on studies of uncomplicated influenza in adults.[6] None of the drugs has been shown to prevent serious complications, such as pneumonia or the exacerbation of underlying disease. Data are limited regarding their effectiveness in the treatment of influenza

Indications for antiviral drugs against influenza
• Patients at risk and their household contacts, who have not (yet) been vaccinated at the time when influenza infections are becoming widespread in an area.
• Control of an outbreak in a (semi-)closed community (e.g. nursing home).
• Patients at risk with a known hypersensitivity to chicken proteins (contraindication for vaccination).
• Vaccine mismatch between vaccine component and circulating virus strain.
• Pandemic threat with no vaccine (yet) available.

Table 14. Indications for antiviral drugs against influenza. Adapted from CDC, Prevention and control of influenza. Recommendations ACIP. *MMWR Recomm Rep* 2004; **53**(RR-6): 1–40.

among people at high risk of serious complications and among paediatric populations.

Both M2 channel blockers interact with a variety of different medicines, which can exacerbate their toxicity, including antihistamines, anticholinergics, co-trimoxazole, triamterene, quinine, quinidine, monoamine oxidase inhibitors, cimetidine, aspirin and paracetamol. When drugs such as antihistamines, antidepressants or minor tranquillizers are co-administered, patients should be monitored closely for CNS side effects.

Adverse effects

The most common side effects of the M2 inhibitors are CNS complaints (anxiety, difficulty concentrating, insomnia, dizziness, headache and jitteriness) and gastrointestinal upset. While the CNS complaints are minor in younger adults, these symptoms may be very prominent in the elderly, particularly with the use of amantadine. Reduction of these symptoms depends on dose adjustments based on creatinine clearance. Patients who receive amantadine may develop antimuscarinic effects, orthostatic hypotension and congestive heart failure. Particularly in the elderly or those with renal failure, serious CNS side effects due to amantadine (and less often rimantadine) include confusion, disorientation, mood alterations, memory disturbances, delusions, nightmares, ataxia, tremors, seizures, coma, acute psychosis, slurred speech, visual disturbances, delirium, oculogyric episodes and hallucinations.[1] Amantadine causes CNS side effects in about 15–30% of people, as well as dose-related abnormalities in psychomotor testing. Amantadine (and possibly rimantadine) may increase the risk of seizures in those with a history of seizures.

In general, NAIs have fewer adverse effects than the M2 channel blockers. Oseltamivir is normally well tolerated but may induce gastrointestinal side effects, including nausea and vomiting, especially if the drug is not taken with food. Other infrequent possible adverse events include insomnia, vertigo and fever. Postmarketing reports suggest that oseltamivir may be associated rarely with skin rash, hepatic dysfunction or thrombocytopenia.

The inhaled zanamivir may induce bronchospasm (potentially severe) in those with underlying lung disease. Current guidelines advise against the use of zanamivir in patients with hyperreactive airways, unless the patient is closely monitored and has a fast-acting inhaled bronchodilator available when inhaling zanamivir.[7] Other less frequent side effects include diarrhoea, nausea, headache and dizziness. Zanamivir is associated with low bioavailability and no clinically significant drug interactions have been recognized.

Influenza prophylaxis with antivirals

As indicated above, under specific conditions, influenza antivirals may be used for prevention of influenza infection. Both of the M2 blockers are 70–90% effective in preventing illness from influenza A infection in healthy adults.[8] When used as prophylaxis, they permit subclinical infection (whilst preventing illness) and allow development of protective antibodies. Both drugs have been studied extensively as a component of influenza outbreak-control programmes and can limit the spread of influenza in nursing homes.[9]

Despite the fact that, of the NAIs, only oseltamivir has been approved for prophylaxis, both drugs have been shown to be similarly effective (82–84%) in preventing influenza in community studies of healthy adults.[10,11] Few data are available regarding their prophylactic use in closed settings or among patients with chronic medical conditions, but one study with oseltamivir reported a 92% reduction in influenza illness among nursing-home residents.[12]

There are no data regarding prophylactic efficacy of any of the four antiviral agents among people with severe immunosuppression. In these cases, prophylaxis can be considered for those expected to have an inadequate antibody response to influenza vaccine, including HIV-positive individuals, in particular those with advanced disease. Such patients should be monitored closely for interactions with other drugs used to treat HIV-positive individuals.

For the purposes of prevention, after an outbreak has been reported, vulnerable adults can be vaccinated and simultaneously receive prophylactic treatment for 2–3 weeks until sufficient antibodies have been generated by vaccination.[13,14] In individuals who remain unvaccinated, the drug must be taken each day for the duration of influenza activity in the community to be maximally effective as prophylaxis. Children under 9 years old receiving the influenza vaccine for the first time can require 6 weeks of prophylaxis, i.e. for 4 weeks after the first dose of vaccine and an additional 2 weeks after the second.

Control of influenza outbreaks in (semi-)closed settings

In closed institutions the use of antiviral drugs both for treatment and prophylaxis is a key component in the control of influenza outbreaks. When an outbreak occurs in a nursing home, all residents should receive prophylaxis, whether or not they have been vaccinated, as vaccination is only ~60% effective in preventing laboratory-confirmed influenza illness among elderly people in this setting.[15] This should continue for a minimum of 2 weeks and until approximately 1 week after the last case has been identified. Prophylaxis should also be considered for all unvaccinated staff, and, in the case of outbreaks caused by a variant strain not well matched by the vaccine, prophylaxis should be considered for all employees, both vaccinated and unvaccinated. The dose of drug should be determined on an individual basis.[16] Prevention can also be considered for controlling outbreaks in, for example, dormitories, schools, cruise ships.

Cost-effectiveness of treatment with antivirals

A recent review article addressed the cost-effectiveness of treatment of ILI with antiviral agents.[17] Treatment with amantadine, zanamivir or oseltamivir was examined in eight studies. Several further studies investigated treatment vs a test-and-treat strategy or vs vaccination. One study included an economic analysis of prophylactic use of antivirals instead of, or on top of, vaccination.[18] Analyses were mostly directed at net costs per symptom-day averted or quality-adjusted life-year (QALY) gained. In some cases net costs were estimated per death averted or life-year gained. A large variation in cost-effectiveness ratios was observed, related to differences in perspectives chosen, populations targeted and diagnostic accuracy assumed. For example, one specific study estimated acceptable incremental net costs per QALY of antivirals compared to standard care at US$18,500, US$28,000 and US$44,000 for amantidine, zanamivir and oseltamivir, respectively, whereas for children, all antivirals exhibited cost-effectiveness over US$150,000 per QALY gained.[18] An analysis of oseltamivir prophylaxis in a Canadian long-term care facility during an influenza A outbreak showed that oseltamivir resulted in less ILI than amantadine prophylaxis, and that oseltamivir prophylaxis saved costs at Can$1,200 per 100 patients compared with amantadine and $3,400 per 100 patients compared with no prophylaxis.[19]

Development of drug resistance

Resistance to amantadine and rimantadine occurs as a result of amino acid substitutions in the transmembrane portion of the M2 protein. Resistant viruses may emerge in patients within 2–4 days after the start of therapy.[20] Although resistant virus was previously uncommon (<1%),[21] recent reports from the CDC in the USA and the Public Health Agency of Canada confirm a rise in the last 2 years from 1.9% to 91% of isolates of influenza A (H3N2) strains that are now resistant to amantidine and rimantidine.* For the 2005–06 influenza season, there is an interim recommendation against the use of amantidine and rimantidine in both countries. As there has been no documented resistance to the NAIs, the interim recommendation advises that oseltamivir (prophylaxis and treatment) and zanamivir (treatment) be selected if an antiviral medication is to be used for the treatment and prophylaxis of influenza. Further, high-risk groups should be given priority for use of these antiviral medications.

*www.cdc.gov/flu/professionals/treatment/0506antiviralguide.htm

Emergence of resistant virus does not appear to cause a rebound in illness in immunocompetent adults, but may be associated with protracted illness and shedding in immunocompromised hosts.[22] Importantly, resistant virus can be spread to others and has caused failures of antiviral prophylaxis under close contact conditions, as in nursing homes[23] and households.[1] The resistant viruses appear to retain wild-type pathogenicity and cause an influenza illness indistinguishable from that caused by susceptible strains.

To date, the use of NAIs for both prophylaxis and treatment has not been associated with clinically relevant development of antiviral resistance during interpandemic use – although, as discussed below, the avian H5N1 virus does appear to develop resistance to oseltamivir in the treatment of H5N1 disease.[24,25] The lack of resistance development of the NAIs often makes them, despite the fact that they are expensive, the preferred drugs relative to the low-cost amantadine and rimantadine, but ultimately cost considerations may also influence the choice of the drug that is used.

The role of antivirals in pandemic control

An essential component of pandemic planning is the use of antiviral drugs early in the pandemic to treat illness and reduce transmission of the virus.[26–28] This would buy time until a vaccine against the pandemic strain can be produced and available in sufficient quantities for mass vaccination. Although a matter of considerable debate, two recent studies, based on simulation of an influenza outbreak in Southeast Asia, have suggested that an emerging pandemic may even be contained altogether at the start with appropriate use of targeted antiviral prophylaxis.[29,30]

Yet, many issues related to the use of antivirals in pandemic control remain. The relatively rapid induction of resistance to the M2 inhibitors when used for the treatment of influenza illness would quickly limit the efficacy of this class of antiviral drugs for preventing influenza illness and ameliorating the impact of pandemic influenza. The novel

NAIs have thus been a welcome addition to the armament to be employed in pandemic control. However, a number of issues related to the use of this class of drugs remains to be addressed even as governments are stockpiling oseltamivir for pandemic preparedness. Recent reports of the development of resistance to oseltamivir during the treatment of H5N1 disease[24,25] raise concerns about the potential for induction and spread of antiviral-resistant H5N1 strains. Issues related to the dependence on a single antiviral drug, stockpiling and how these drugs will be distributed, and indications for the prevention and treatment of pandemic influenza remain to be clarified. In addition, the efficacy of these drugs in the absence of vaccination in the very young[16] and the very old[31] also remain in question in both pandemic and interpandemic influenza seasons.

The need for large stockpiles of lower cost, more efficacious and easily administered medications has been identified.[27] Plans could include the use of antivirals to treat severe illness during a pandemic and prophylax all health-care and related support workers, but this plan may have limited feasibility from a governmental perspective.[32] Governments are attempting to stockpile oseltamivir to be used in an influenza pandemic, but who would receive the drug when supplies are limited? Also, the potential for criminal activities around the limited supplies is a significant concern. Drug distribution issues, including central vs local stockpiles, need to be carefully considered. Multiple drug strategies for prophylaxis and treatment indications could include the use of zanamivir (inhaled drug formulation) and potentially amantidine and rimantidine or generic formulations of the neuraminidase inhibitors where drug supply and cost are the major barriers to pandemic preparedness. Co-ordinated public health planning and the commitment of governments to the development of realistic plans for including antivirals in pandemic planning are necessary.[26–28]

References

1. Hayden FG, Aoki FY. Amantadine, rimantadine, and related agents. In: Barriere SL, editor. *Antimicrobial Therapy and Vaccines*. Baltimore: Williams & Wilkins; 1999; pp. 1344–1365.

2. Colman PM. Influenza virus neuraminidase: structure, antibodies, and inhibitors. *Protein Sci* 1994; **3**: 1687–1696.

3. Wang C, Takeuchi K, Pinto LH, Lamb RA. Ion channel activity of influenza A virus M2 protein: characterization of the amantadine block. *J Virol* 1993; **67**: 5585–5594.

4. Ruigrok RW, Hirst EM, Hay AJ. The specific inhibition of influenza A virus maturation by amantadine: an electron microscopic examination. *J Gen Virol* 1991; **72**: 191–194.

5. National Institute for Clinical Excellence. Guidance on the use of zanamivir, oseltamivir and amantadine for the treatment of influenza. In: *Flu – zanamivir (review), amantadine and oseltamivir*. 2003; **58**. (www.nice.org.uk)

6. Food and Drug Administration. Subject: safe and appropriate use of influenza drugs [Public Health Advisory]. Rockville, MD: US Department of Health and Human Services, Food and Drug Administration, 2000.

7. Prevention and control of influenza: Recommendations of the Advisory Committee on Immunization Practices (ACIP). *MMWR* 2002; **51**: 1.

8. Demicheli V, Jefferson T, Rivetti D, Deeks J. Prevention and early treatment of influenza in healthy adults. *Vaccine* 2000; **18**: 957–1003.

9. Nicholson KG. Use of antivirals in influenza in the elderly: prophylaxis and therapy. *Gerontology* 1996; **42**: 280–289.

10. Monto AS, Pichichero ME, Blanckenberg SJ *et al*. Zanamivir prophylaxis: an effective strategy for the prevention of influenza types A and B within households. *J Infect Dis* 2002; **186**: 1582–1588.

11. Hayden FG, Atmar RL, Schilling M *et al*. Use of the selective oral neuraminidase inhibitor oseltamivir to prevent influenza. *New Engl J Med* 1999; **341**: 1336–1343.

12. Peters PH Jr, Gravenstein S, Norwood P *et al.* Long-term use of oseltamivir for the prophylaxis of influenza in a vaccinated frail older population. *J Am Geriatr Soc* 2001; **49**: 1025–1031.

13. Gross PA, Russo C, Dran S, Cataruozolo P, Munk G, Lancey SC. Time to earliest peak serum antibody response to influenza vaccine in the elderly. *Clin Diagn Lab Immunol* 1997; **4**: 491–492.

14. Brokstad KA, Cox RJ, Olofsson J, Jonsson R, Haaheim LR. Parenteral influenza vaccination induces a rapid systemic and local immune response. *J Infect Dis* 1995; **171**: 198–203.

15. Guy R, Lambert S, Kelly H. Estimating influenza vaccine effectiveness in an outbreak when anti-viral medications were used as a control measure. *Aust NZ J Public Health* 2005; **29**: 540–543.

16. Dolin R. Influenza – interpandemic as well as pandemic disease. *New Engl J Med* 2005; **353**: 2535–2537.

17. Lynd LD, Goeree R, O'Brien BJ. Antiviral agents for influenza; a comparison of cost-effectiveness data. *Pharmacoeconomics* 2005; **23**: 1083–1106.

18.Turner D, Wailoo A, Nicholson A *et al.* Systematic review and economic decision modelling for the prevention and treatment of influenza A and B. National Institute of Clinical Excellence (NICE), Leicester (UK), 2002.

19. Risebrough NA, Bowles SK, Simor AE *et al.* Economic evaluation of oseltamivir phosphate for postexposure prophylaxis of influenza in long-term care facilities. *J Am Geriatr Soc* 2005; **53**: 444–451.

20. Hayden FG. Amantadine and rimantadine: clinical aspects. In: Richman DD, editor. *Antiviral Drug Resistance*. New York: Wiley; 1996.

21. Ziegler T, Hemphill ML, Ziegler ML *et al.* Low incidence of rimantadine resistance in field isolates of influenza A viruses. *J Infect Dis* 1999; **180**: 935–939.

22. Englund JA, Champlin RE, Wyde PR *et al.* Common emergence of amantadine- and rimantadine-resistant influenza A viruses in symptomatic immunocompromised adults. *Clin Infect Dis* 1998; **26**:1418–1424.

23. Gravenstein S, Drinka P, Osterweil D *et al.* Inhaled zanamivir versus rimantadine for the control of influenza in a highly vaccinated long-term care population. *J Am Med Dir Assoc* 2005; **6**: 359–366.

24. De Jong MD, Tran TT, Truong HK *et al.* Oseltamivir resistance during treatment of influenza A (H5N1) infection. *New Engl J Med* 2005; **353**: 2667–2672.

25. Le QM, Kiso M, Someya K *et al.* Avian flu: isolation of drug-resistant H5N1 virus. *Nature* 2005; **437**: 754.

26. Monto AS. The role of antivirals in the control of influenza. *Vaccine* 2003; **21**: 1796–1800.

27. Stohr K. Preventing and treating influenza. *Br Med J* 2003; **326**: 1223–1224.

28. Monto AS. Vaccines and antiviral drugs in pandemic preparedness. *Emerg Inf Dis* 2006; **12**: 55–60.

29. Ferguson NM, Cummings DA, Cauchemez S *et al.* Strategies for containing an emerging influenza pandemic in Southeast Asia. *Nature* 2005; **437**:209–214.

30. Longini IM, Halloran ME, Nizam A, Yang Y. Containing pandemic influenza with antiviral agents. *Am J Epidemiol* 2004; **159**: 623–633.

31. Ambrozaitis A, Gravenstein S, van Essen GA *et al.* Inhaled zanamivir versus placebo for the prevention of influenza outbreaks in an unvaccinated long-term care population. *J Am Med Dir Assoc* 2005; **6**: 367–374.

32. Cinti S, Chenoweth C, Monto AS. Preparing for pandemic influenza: should hospitals stockpile oseltamivir? *Infect Control Hosp Epidemiol* 2005; **26**: 852–854.

8. Vaccination: Cornerstone of Influenza Control

Vaccination is the primary and single most cost-effective method of preventing influenza. Flu vaccine development began just a few years after the first isolation of the influenza virus in 1933.[1,2] Pioneering studies demonstrated that influenza A/PR/8/34 (H1N1) virus would infect humans upon subcutaneous administration, inducing virus-neutralizing antibodies.[2,3] Soon after these initial observations, studies using formalin-inactivated whole-virus preparations were conducted, the first inactivated influenza vaccines being introduced in the 1940s.[2] Most current influenza vaccines are also inactivated formulations, consisting of either split virus or subunit preparations, the latter containing just the isolated viral haemagglutinin (HA) and neuraminidase (NA).[2,4,5] These vaccines are generally produced from virus grown on embryonated chicken eggs.

Active immunization against any infectious disease, including influenza, aims at induction of antimicrobial immunity by inoculating the person with an attenuated or inactivated form of the pathogen involved. Following such an approach, one attempts to closely mimic the immune response to a natural infection, which is often considered the "gold standard" for protection. Thus far, the hallmark for influenza vaccine efficacy has been the induction of an adequate level of virus-neutralizing antibodies in the serum.[5–7] These antibodies are primarily directed against the envelope glycoproteins of the virus, HA and NA, HA being the major target for virus-neutralizing antibodies (see Chapter 4). It is well established that HA-specific antibodies in the circulation protect from severe viral pneumonia as a result of transudation of these antibodies from the blood into the lungs.[6,7]

Current influenza vaccines contain antigens from two influenza A virus strains and one B strain, according to annual recommendation by the WHO.[8,9] To ensure an optimal antigenic match between the virus strains in the vaccine and the viruses circulating in the subsequent influenza season, this WHO recommendation is based on intensive surveillance of new influenza strains around the globe, allowing an informed selection of strains to be included in the next annual vaccine.[8,9]

Today, many countries have implemented influenza vaccination programmes, the primary target groups for vaccination including the elderly and people with underlying medical conditions which make them vulnerable to serious complications of influenza.[9–11] Also, there is increasing awareness of the potential societal and health benefits of vaccinating working adults and children.[12,13]

The safety record of inactivated influenza vaccines is excellent.[9,14] Hundreds of millions of vaccine doses are distributed worldwide each year, adverse effects being extremely rare. Influenza vaccination has been shown to be highly effective. Vaccination results in reductions of influenza-related respiratory illness and numbers of physician visits among all age groups, and in lower hospitalization rates and deaths among the elderly and patients at risk for serious complications of influenza.[12,13] Vaccination coverage among target groups has increased considerably in recent years[10] as the awareness of the impact of influenza is growing and influenza has become an important issue on the public-health agenda in many countries.[15,16] However, the use of available influenza vaccines is still far from optimal. Health-care professionals are in a key position to explain the favourable risk–benefit ratio of influenza vaccination to people in target groups and thus to motivate them to take the vaccine. Indeed, increased use of influenza vaccines is expected to significantly reduce epidemics and to improve our preparedness for potential new pandemic outbreaks.

Key Messages

- Current inactivated influenza vaccines have an excellent safety record.
- Influenza vaccination results in significant reductions in influenza-related respiratory illness, hospitalization rates and deaths among the elderly and patients at risk for serious complications of influenza.
- Influenza vaccination of the elderly and patients with underlying medical conditions is cost-effective and in many cases cost-saving.
- Scientific evidence also demonstrates the benefits of vaccinating healthy working adults and children.
- Misconceptions about the benefits of influenza vaccination and overestimation of its risks lead to decreased acceptance and suboptimal use of influenza vaccines.
- Primary-care physicians and other health-care workers play a pivotal role in explaining the favourable risk–benefit ratio of influenza vaccination and in motivating patients in target groups to take the vaccinaton.
- The WHO strongly encourages the use of influenza vaccine among the elderly and people in other target groups.

Current inactivated influenza vaccines

Vaccine formulations

Initial inactivated influenza vaccine formulations invariably consisted of whole inactivated virus formulations.[2,4] Whole inactivated virus vaccines are generally quite immunogenic. However, despite the introduction of improved techniques for virus purification, local reactogenicity and systemic side effects remain a problem associated with the use of these vaccines, particularly in small children.[17] This has led to the development of vaccine formulations consisting of disrupted virus particles, which turned out to be almost equally immunogenic in primed individuals, yet causing significantly fewer side effects.[2] These split-virus vaccines, first licensed in the USA in 1968, are among the most widely used formulations to date. While ether was the original splitting agent of choice, currently detergents such as Tween 80, Triton N101, cetyl trimethyl ammonium bromide (CTAB) and sodium deoxycholate are used in addition to ether to disrupt the virus particles.[4] A disadvantage of split vaccines, relative to whole inactivated virus, is their comparatively low immunogenicity in unprimed individuals, such as young children without prior exposure to flu virus or vaccine, requiring a booster immunization for adequate protection.[18]

Relative to split-virus vaccines, subunit preparations represent a further refinement and improvement of the vaccine formulation. Subunit vaccines contain the HA and NA surface glycoproteins purified from other viral components.[4] Because of their high purity, subunit vaccines have a favourable profile in terms of local and systemic side effects, compared to whole-virus and split vaccines.[19] Yet, subunit vaccines are equally immunogenic in primed individuals. The concept of using just the isolated viral HA and NA antigens is based on the notion that the primary correlate of protection against influenza is an adequate level of circulating antibodies against the surface glycoproteins of the virus, especially HA (see Chapter 4). Subunit antigens are isolated by detergent solubilization of the viral envelope, followed by sedimentation of the nucleocapsid through ultracentrifugation. Removal of the detergent from the supernatant by adsorption onto a hydrophobic resin, such as Amberlite, results in the formation of rosettes of HA and NA, with only small amounts of contaminating core proteins, such as NP or M1, and viral membrane lipid.[4] Influenza subunit vaccines were first licensed in UK in 1980 and are now used in many countries worldwide.

Selection of vaccine strains

To be effective, the vaccine components need to match those of the circulating influenza virus strains in the target season (see Chapter 3). Current influenza vaccines contain three virus strains, two A strains (an H3N2 and an H1N1 strain) and one B strain. These strains are included in the vaccines on recommendation of the WHO.[8,9] This recommendation is based on an extensive review of epidemiological data and antigenic and genetic analyses of virus isolates by the four WHO Collaborating Centres. Recently, a mathematical modeling approach for mapping of the antigenic distance between virus strains has been included in the strain selection process.[20] In order to allow vaccine manufacturers sufficient time for production, in February of each year the WHO issues its recommendation about which viral strains should be included in the next winter's vaccine for the northern hemisphere. A second review follows in September to consider adjustments of the vaccine composition for the southern hemisphere.

That this system of surveillance and recommendation works quite well is demonstrated by the good match achieved in, for example, the influenza seasons from 1987 to 1997.[5] Within this period, 23 vaccine strains recommended by the WHO matched with the subsequently circulating total of 30 virus strains (Table 15). A complete match of all three strains was achieved in five out of 10 seasons. Nevertheless, an intrinsic uncertainty remains and sometimes there is not a perfect match between vaccine strains and circulating viruses, in which case the vaccine may have reduced efficacy. However, these occasional mismatches should by no means be regarded as a general justification for not providing or taking the vaccination.

Dose standardization

The development of the single radial immunodiffusion (SRD) test has allowed the implementation of a stringent standardization of vaccine dosaging.[5] The SRD test determines the content of HA antigen in influenza vaccines on the basis of the immunological activity of the antigen against a reference sheep antiserum (Figure 24).[5] The application of this assay and the use of standardized reagents, supplied by one of the WHO Collaborating Centres to influenza vaccine manufacturers, thus guarantee

Match between vaccine recommendations and concurrent epidemic influenza virus strains			
Season	A/H1N1	A/H3N2	B
1987–1988	+	−	−
1988–1989	+	+	+
1989–1990	+	+	+
1990–1991	+	−	+
1991–1992	+	+	+
1992–1993	+	−	+
1993–1994	+	−	+
1994–1995	+	−	−
1995–1996	+	+	+
1996–1997	+	+	+

Table 15. Match between WHO vaccine recommendations and epidemic virus strains circulating in subsequent winter season. Adapted from Wood JM. Standardization of inactivated influenza vaccines. In: Nicholson KG, Webster RG, Hay AJ, editors. *Textbook of Influenza*. Blackwell Science, 1998; pp. 333–345[5] with permission from Blackwell Publishing.

optimal uniformity among different influenza vaccine formulations in terms of potency.[5]

Several studies have shown that a vaccine dose of ≥10 μg HA per strain induces an adequate immune response in primed individuals.[21,22] Current trivalent inactivated influenza vaccines contain 15 μg of HA per strain as assessed by the SRD assay. This dose scheme has been formally standardized.[5,23,24] The trivalent inactivated influenza vaccine is administered by intramuscular or deep subcutaneous injection.

Evaluation of vaccine immunogenicity

Not only has the composition of influenza vaccines been standardized in many countries, in the EU, criteria for

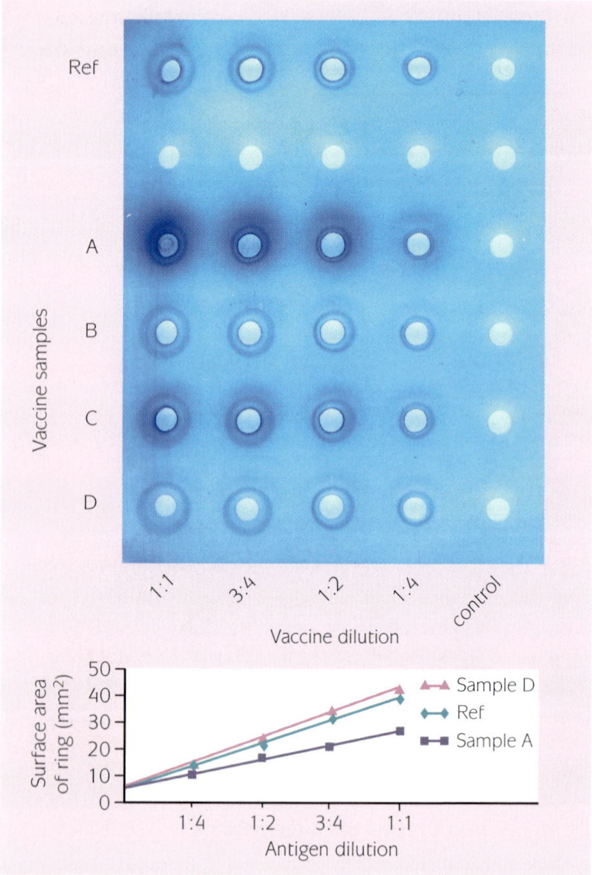

Figure 24. Single radial immunodiffusion (SRD) test of influenza vaccine potency.[5] Serial dilutions of detergent-treated vaccine, and a reference antigen (Ref), are added to wells in an agarose gel containing a sheep antiserum against the relevant HA. The surface area of the precipitation rings formed (top) is subsequently plotted as a function of the dilution of the vaccine (bottom). The slope of the resulting curves is a direct measure of the HA concentration in the vaccine, and is compared with that of the reference curve. In the graph, data for vaccine samples A and D and the reference antigen shown in the top gel are plotted. Courtesy of Jeroen Medema, Solvay Pharmaceuticals, Weesp, the Netherlands.

vaccine immunogenicity have also been implemented.[5,22,23] As indicated above, current inactivated influenza vaccines aim at induction of an efficient systemic antibody response against the viral surface antigens. Accordingly, vaccine efficacy is evaluated on the basis of serological data, in particular antibody titres and seroprotection rates.

Antibody titres are generally determined on the basis of haemagglutination–inhibition (HI) activity[5,6] or, occasionally, by single radial haemolysis (SRH).[5] In the HI assay, serum samples of vaccinated individuals are tested for their ability to inhibit agglutination of erythrocytes, induced by influenza virus through interaction of the viral HA with sialic acid on the red cell surface. Antibodies directed against HA interfere with this interaction, thus

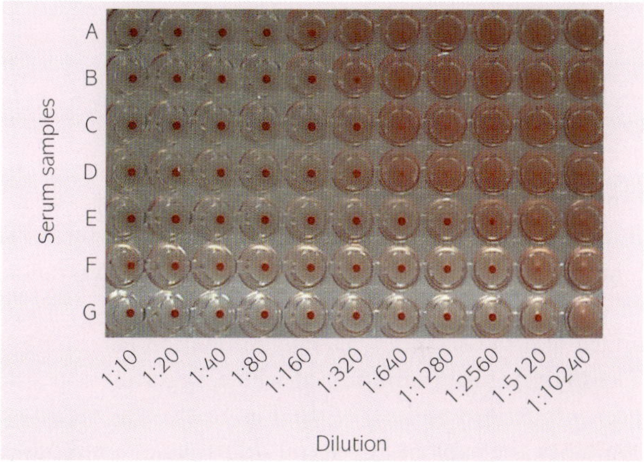

Figure 25. Haemagglutination–inhibition (HI) titration of an antiserum against influenza. Turkey or guinea pig erythrocytes are incubated with a standard amount of influenza virus and two-fold dilutions of the serum to be evaluated. The reciprocal of the highest dilution at which haemagglutination is completely inhibited (haemagglutination inhibition is observed as the formation of a small concentrated dot of red cells in the bottom of the well) is defined as the HI titre of the sample. For example, the HI titre of serum C is 320, that of serum F is 2560. Courtesy of René Benne, Laboratory of Infectious Diseases, Groningen, the Netherlands.

European Union criteria for the assessment of vaccines		
Criterion	**18–60 years**	**>60 years**
Seroconversions or significant rises in anti-HA antibody titre	>40%	>30%
Mean geometric increase in titre	>2.5	>2.0
Patients achieving HI titre ≥ 40 or SRH titre >25 mm^2	>70%	>60%
For each virus strain, at least one of the above criteria should be met		

Table 16. Criteria of the European Medicines Agency (EMEA) of the EU for the evaluation of influenza vaccine efficacy. Issued by the Committee on Proprietary Medicinal Products (CPMP). Note for guidance on harmonization of requirements for influenza vaccines. CHMP/BWP/214/96, 1997 (www.emea.eu.int/pdfs/human/bwp/021496en.pdf)[23] © EMEA 1997 Reproduction and/or distribution of this document is authorized for non-commercial purposes only provided the EMEA is acknowledged.

inhibiting the agglutination process. In practice, serial dilutions of serum are incubated with a standardized amount of influenza virus, after which a fixed concentration of erythrocytes, generally from guinea-pigs or turkeys, is added and the extent of haemagglutination determined. The highest serum dilution at which agglutination still occurs is defined as the HI titre (Figure 25). The SRH test evaluates the capacity of serial dilutions of antibodies against HA to lyse erythrocytes in the presence of guinea-pig complement.[5]

The European Medicines Agency (EMEA) has formally standardized the EU requirements for annual evaluation of influenza vaccine efficacy (Table 16).[23,24] These requirements include specified numbers of seroconversions and/or numbers of individuals achieving a certain antibody titre upon vaccination within a study population of a specified age range. In addition, there is a requirement for an annual clinical study among 50 volunteers of 18–60 years old and 50 volunteers of age >60 years. However, after evaluation of the long-term experience with the EU-specific registration process, the value and need for these annual serological studies for relicensure of influenza vaccines have recently been questioned.[25]

Annual timetable for vaccine production and licensing

As almost all current influenza vaccines are prepared from egg-grown virus (Figure 26), the annual vaccine production cycle begins with the estimation and ordering of the required numbers of embryonated chicken eggs well before actual vaccine production starts.[26] Then, after the WHO has issued its recommendation, seed viruses are generated and characterized for approval by the WHO Collaborating Centres. These seed viruses are high-growth reassortants (see Chapter 3), carrying the HA and NA of the recommended vaccine virus on the background of a virus that replicates well in embryonated eggs. Generally, 1–2

Figure 26. Production of influenza vaccine virus on embryonated chicken eggs. Courtesy of Solvay Pharmaceuticals, Weesp, the Netherlands.

months after the WHO recommendation, seed virus lots are released to the manufacturers. Vaccine production then continues for several months. In this process, quantification of the potency of monovalent vaccine bulks is a critical step, requiring the availability of specific reagents for SRD determination. Finally, the ultimate trivalent vaccine can be formulated and the syringes or vials filled. Immediately after the first batches of trivalent vaccine have been released for use, two serological clinical studies (in people aged 18–60 and >60 years) are performed to satisfy the licensing criteria of the European regulatory authorities,[23,24] as discussed above. Results of these clinical studies form part of the annual registration dossier for market authorization in the member states of the EU.

From the time of recommendation by the WHO of virus strains to be included in the vaccine, it takes approximately 6 months to bring the vaccine to the market. Within two campaigns for the northern and southern hemispheres, a total of about 300 million doses of influenza vaccine are currently being produced each year.

Vaccination coverage

For a long time, doubts and misconceptions about the risk–benefit ratio of influenza vaccination have hampered the implementation of recommended policies for influenza vaccination. However, in recent years, influenza vaccination has become a prominent issue on the public-health agenda in an increasing number of countries.[10] Many developed as well as developing countries have now adopted formal recommendations on influenza vaccination for specific target groups. Although these recommendations are still a matter of national responsibility, they have nonetheless become increasingly uniform, particularly throughout Western Europe, North America, Australia, New Zealand and Japan. In addition, in 2000, the WHO has issued for the first time a formal recommendation for the use of inactivated vaccines for the prevention of influenza.[9]

Recommendations for influenza vaccination

Table 17 presents the recommendations for influenza vaccination adopted in most countries. The elderly represent the primary target group.[10,11] This recommendation follows the increased susceptibility of the elderly for infectious diseases in general, which may be explained, at least partly, by a gradual decline in immune competence with age, particularly at the level of T cell function (see Chapter 4). The recommendation is also based on the proven clinical

Influenza vaccination recommendations in European countries
• People aged >65 years.
• Residents of nursing homes and other long-term care facilities.
• Adults and children with chronic pulmonary disorders.
• Adults and children with chronic cardiovascular disorders.
• Those who have required regular medical follow-up or hospitalization during the preceding year because of – chronic metabolic diseases (including diabetes mellitus), – renal dysfunction, – haemoglobinopathies, or – immunosuppression (including immunosuppression caused by medications or by human immunodeficiency virus).
• Children and teenagers (aged 6 months to 18 years) who are receiving long-term aspirin therapy and therefore might be at risk for developing Reye's syndrome after influenza infection.
• Vaccination of health-care workers and others in close contact with persons at high risk, including household members, is recommended.

Table 17. Recommendations for influenza vaccination of specific target groups adopted in most European countries. Source: The Macroepidemiology of Influenza Vaccination (MIV) Study Group. The macroepidemiology of influenza vaccination in 56 countries, 1997–2003. *Vaccine* 2005; **23**: 5133–5143.[10]

efficacy and effectiveness of flu vaccination of the elderly.[12] In most countries, flu vaccination is recommended for all individuals above 60 or 65 years of age. In a few countries, including the USA and Belgium, the age recommendation includes everyone over the age of 50 years due to the increased prevalence of high-risk conditions for influenza illness among people over 50 years of age. In general, an age-based strategy is easier to implement than a targeted risk-group strategy. This is particularly relevant because of the relatively low vaccine coverage rates in patients at risk who are younger than 65 years of age.[27–29] Currently, more countries are considering lowering the age limit for vaccination recommendation. In recent studies, the cost-effectiveness of an expanded vaccination policy, including all individuals over 50 years of age, has been established for a number of European countries.[30,31] Vaccination is also quite uniformly recommended for individuals with specific medical conditions, including cardiovascular, pulmonary and renal disease, diabetes and immunodeficiency.[10] Furthermore, many countries recommend flu vaccination for nursing-home residents, health-care workers and household contacts of high-risk individuals.[10] In addition, there is an increasing awareness of the potential benefits of influenza vaccination of younger healthy (working) adults,[12,13] pregnant women,[32] and children.[12,33–35]

Current use of influenza vaccines

Along with the implementation of formal recommendations for flu vaccination in an increasing number of countries, the use of influenza vaccines has increased considerably in the last decade or so. The most dramatic changes in this respect have occurred in Korea, Latin America, Japan and some

Figure 27. Use of influenza vaccine in 56 developed and developing countries in the period 1997–2003. Reproduced from The Macroepidemiology of Influenza Vaccination (MIV) Study Group. The macroepidemiology of influenza vaccination in 56 countries, 1997–2003. *Vaccine* 2005; **23**: 5133–5143[10] with permission from Elsevier.

Canada 344
Rep. of Korea 311
United States 286
Japan 230
Australia 214
Spain 212
Germany 210
Italy 206
UK 200
Hong Kong SAR 191
The Netherlands 191
Iceland 181
Belgium 177
New Zealand 173
France 171
Switzerland 170
Greece 165
Ireland 148
Portugal 145
Taiwan 145
Israel 144
Finland 136
Hungary 135
Denmark 133
Croatia 130
Austria 127
Sweden 127
Chile 126
Uruguay 105
Norway 102
Slovenia 99
Russian Fed. 99
Slovak Rep. 98
Singapore 90
Brazil 87
Argentina 85
Poland 79
Czech Rep. 73
Romania 67
Latvia 65
Lithuania 53
Costa Rica 49
Bulgaria 46
Estonia 40
Mexico 29
South Africa 25
Turkey 19
Ukraine 18
Jordan 9
Uzbekistan 6
Albania 2
Georgia 2
Moldova 2
Egypt 1
Kyrgyz Rep. 1
Thailand 1
1997
2003
0 50 100 150 200 250 300 350 400
Vaccine doses distributed/1000 population

central and eastern European countries.[10] Figure 27 presents a survey of influenza vaccine distribution in 56 developed and rapidly developing countries in 1997 and 2003.

As mentioned above, the total number of influenza vaccine doses distributed annually worldwide has increased in recent years to an estimated level of 300 million in 2003. About 60% of these doses are distributed among North America and Western Europe and 40% goes to the rest of the world. In 1994, these figures were 80% and 20%, respectively.[10] This trend indicates that many countries, including developing countries, are moving towards implementation of measures for influenza prevention and control on an annual basis.

However, despite the increased vaccine use, recent surveys in Europe show that the coverage rates in target populations are still far from the WHO-recommended 75% in 2010.[29,36] The current coverage rates range from 18% (Poland) to 67% (Spain) for the elderly and from 3% to 40% for various risk groups in younger populations.[28,29,37] In the USA, only 35% of adults between the ages of 18 and 64 years who are at risk for serious complications due to influenza were being vaccinated in 2003.[38] The implication of these findings is that many elderly and at-risk patients are not receiving the best possible protective treatment to prevent influenza or minimize the consequences of an influenza infection. Based on current scientific evidence of vaccine efficacy and effectiveness,[12] serious efforts are to be made towards compliance with the target vaccination rates recommended by the WHO for 2010.[36]

The single most important factor influencing the use of influenza vaccine is whether it is recommended by the doctor. Therefore, primary-care physicians and other health-care workers play a major role in implementing influenza vaccination programmes,[29] as discussed in more detail in the last paragraph of this chapter.

Benefits of influenza vaccination

The value of immunization against influenza is sometimes being questioned, as outbreaks of flu continue despite increased influenza vaccination coverage.[39] Therefore, there has been a strong demand for sound scientific data on the effects of influenza vaccination. As a result, many clinical studies have now produced consistent data showing the clear-cut benefits of influenza vaccination.[12,13,33,35,40] Since the elderly comprise by far the largest target population for flu vaccination, the majority of studies evaluating the benefits of vaccination have been conducted among people in this age group; these will be discussed in more detail below. Individual studies have also indicated clear-cut beneficial effects in other target groups, such as diabetics[37] and patients with asthma[41] or cardiovascular disease.[42]

Vaccine efficacy and clinical effectiveness of influenza vaccination

In evaluating the outcome of influenza vaccination, a distinction is often made between vaccine efficacy *per se* and the clinical effectiveness of vaccination.[12,13] Vaccine efficacy is defined as the reduction in the rate of laboratory-confirmed influenza among vaccinated compared to non-vaccinated individuals. Vaccine efficacy thus provides a direct measure of the specific reduction in influenza infection rates as a result of the vaccination. With an approximate annual attack rate of influenza of 5–10%, accurate determination of vaccine efficacy requires well-controlled studies and evaluation of large study groups. Obviously, vaccine efficacy will depend critically on the match between the vaccine composition and the circulating virus strains.

Clinical effectiveness provides a less specific, yet quite relevant and important, measure of the benefit of influenza vaccination. It is defined as the reduction of clinically relevant, but not necessarily influenza-specific, disease in a "real-life" situation, including all influenza-like illness (ILI), hospitalizations due to pneumonia from all causes or death from all causes.[12,13,40] As this parameter includes – by definition – disease that is not caused by the influenza virus, clinical effectiveness of vaccination is generally

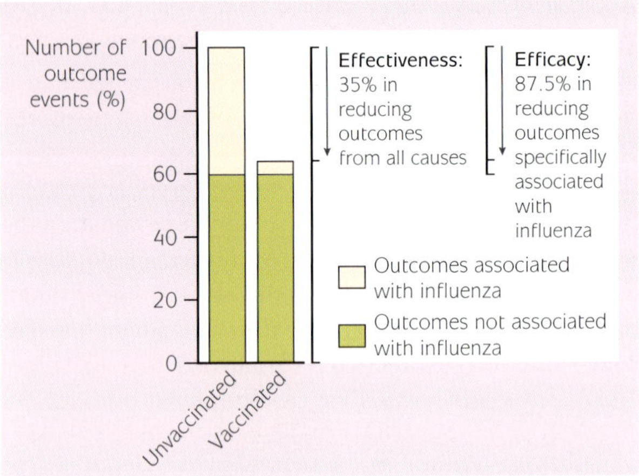

Figure 28. Relationship between influenza vaccine efficacy and clinical effectiveness of influenza vaccination. The figure shows a hypothetical example in which vaccination is associated with a 35% reduction in all outcomes evaluated (such as hospitalizations for pneumonia). However, not all outcomes are due to influenza. If only 40% of the outcomes represented complications of influenza, the underlying efficacy of the vaccine preventing direct influenza-associated outcomes would be 35%/0.4 = 87.5%. Adapted from Nichol KL. Efficacy/clinical effectiveness of inactivated influenza virus vaccines in adults. In: Nicholson KG, Webster RG, Hay AJ, editors. *Textbook of Influenza*. Blackwell Science, 1998; pp. 358–372[13] with permission from Blackwell Publishing.

estimated to be lower than the actual vaccine efficacy, as illustrated by the hypothetical example presented in Figure 28.[13] Therefore, clinical effectiveness should not be confused for vaccine efficacy, as this may result in a substantial underestimation of the actual performance of the vaccine.

Because of the non-specific criteria of influenza-like illness (ILI), the outcome of clinical effectiveness studies depends on the used case definition in the particular study. Indeed, applying different case definitions to one data set from a single clinical study resulted in very different study outcomes.[43] A recent literature review revealed that all published clinical effectiveness studies applied different case definitions for ILI[44] and that by simply changing the sensitivity and specificity of case definitions and observation periods, differences of as much as 30% in study outcome for vaccination effectiveness may be found. Therefore, it was recommended that internationally agreed case definitions for ILI should be developed, as well as observational periods for future influenza clinical effectiveness studies. In recent systematic reviews (Cochrane) on influenza vaccination effectiveness in healthy adults[45] and children,[46] literature data were pooled for the review process, but the differences in case definitions of individual studies were not taken into account. Because of this methodological problem, interpretation of such combined vaccination effectiveness estimates should be made with caution.[34,44] Clearly, these considerations are also relevant for health-economic analyses of influenza vaccination programmes.[47]

Health benefits of vaccination of the elderly

Several studies have indicated that, in cases of a good match between vaccine strain and circulating virus, the efficacy of current inactivated influenza vaccines among the elderly is approximately 60%. For example, in a large, randomized, double-blind, placebo-controlled trial among 1838 subjects of 60 years of age or older in the Netherlands, vaccine efficacy was found to be 58%.[48] This trial was conducted in the 1991–92 winter season and involved the use of a multivalent inactivated influenza vaccine, matching well with the circulating virus.

Numerous studies have convincingly demonstrated the clinical benefits of influenza vaccination in the elderly.[12,13,40,48,49] For example, in a large study in the USA, spanning two influenza seasons (1998–2000) and involving 300,000 community-dwelling elderly people (≥65 years), influenza vaccination was performed in 55.5–59.7% of the population. The vaccination was associated with significant

Clinical effectiveness of influenza vaccination of the elderly	
Outcome measure	**Reduction**
Community-dwelling senior citizens	
Hospitalizations for	
Pneumonia from all causes	33%
All respiratory conditions	32%
Congestive heart failure	27%
Death from all causes	50%
Elderly in nursing homes	
Respiratory illness	56%
Pneumonia from all causes	53%
Hospitalization in general	48%
Death from all causes	68%

Table 18. Clinical effectiveness of influenza vaccination of the elderly. Adapted from Nichol KL. The efficacy, effectiveness and cost-effectiveness of inactivated influenza virus vaccines. *Vaccine* 2003; **21**: 1769–1775[12] with permission from Elsevier.

reductions in pneumonia (29–32%), cardiac disease (19%) and cerebrovascular disease (16–23%).[42] Other studies have indicated that influenza vaccination of patients who have had a myocardial infarction results in a significant reduction in 1-year mortality rates (66% reduction) or risk of further ischaemic events. A meta-analysis, including a large number of individual studies among senior citizens living in the community, concluded that vaccination significantly reduces hospitalization and death rates among the elderly (Table 18).[50] Another meta-analysis has shown that influenza vaccination is also highly effective among residents of nursing homes (Table 18).[51] These findings necessitate a proactive immunization practice by health-care providers in order to allow more elderly people to benefit from existing safe and efficacious influenza vaccines.

Cost-effectiveness of vaccination of the elderly

Economic evaluations, conducted in many different countries, have indicated that vaccination of senior citizens against influenza is always cost-effective and frequently cost-saving.[12,13] For example, in a 6-year study carried out in Minnesota, USA, influenza vaccination of nursing-home residents was associated with an average net saving of $73 per person as a result of reductions in direct medical costs.[12] Vaccination appears to be cost-effective or even cost-saving for both healthy senior citizens and high-risk elderly with underlying chronic medical conditions. In a study conducted in the Netherlands in the 1995–96 and 1997–98 seasons, influenza vaccination was found to be cost-saving for high-risk elderly and cost-effective for all elderly and elderly at low risk, the cost-effectiveness ratios being €1820 per life-year gained for all elderly and €6900 per life-year gained for those at low risk.[52] Similar studies have been conducted in other countries with similar outcomes.

Vaccine safety and contraindications

Inactivated influenza vaccines have an excellent safety record.[9,14] Currently, about 300 million vaccine doses are being administered annually around the globe,[10] and the overall rate of adverse reactions is extremely low. However, sporadically, adverse reactions do occur. The most frequently occurring side effects are local reactions at the site of injection, which usually do not last more than 1–2 days.[14] Generally, the reactions are mild and of a transient nature. It has also been demonstrated that influenza vaccination of patients with asthma is safe.[53] In a large, double-blind, placebo-controlled study, it was demonstrated that no exacerbation of asthma in children occurred as a result of vaccination.[54]

It is important to note that there is no vaccination procedure or, for that matter, any medical intervention that bears no risks whatsoever. However, clearly, the benefits of influenza vaccination by far outweigh its risks. Nevertheless, there are widespread misconceptions regarding the risk–benefit ratio of influenza vaccination. Indeed, some vaccinated individuals even complain of having acquired "the flu" as a result of the vaccination. Primary-care physicians and other health-care workers have the responsibility to explain to their patients the favourable risk–benefit ratio of influenza vaccination.

In specific subpopulations, influenza vaccination is contraindicated. These include people with hypersensitivity to eggs and/or a history of immediate allergic reactions following the vaccination. Allergic reactions to flu vaccination are most likely due to the presence of traces of egg-derived components in the vaccine, ovalbumin in particular. Novel developments, specifically the production of influenza vaccine by cell culture technology (see Chapter 9), should eliminate this problem.

There have been reports about a possible association between GBS and influenza vaccination, particularly during the swine flu vaccination campaign in the USA in 1976–77.[55] More recent studies suggest that GBS may occur at a very low rate of about one additional case per million vaccinees.[56] However, it is unclear whether this marginal increase, against a background incidence of GBS of 10–20 per million, is truly associated with influenza vaccination.

How may influenza control be further improved?

Improvement of vaccine use in target groups

As indicated above, vaccination rates among target populations are not in line with recommendations by WHO,[10,11,36] despite the well-established efficacy and effectiveness of current inactivated influenza vaccines.[12,13,33,40,48–51] Clearly, influenza vaccines are still seriously underutilized.[10]

The underutilization of influenza vaccines is often due to perceptions related to influenza and influenza vaccination which are based on insufficient or inappropriate information. All too often influenza is viewed as a comparatively mild disease that does not pose a serious

Possible actions by primary-care physicians to encourage vaccine uptake in target populations
• Mark and update the records of people recommended for vaccination.
• Send invitation letters together with information leaflets to people recommended for vaccination.
• Organize vaccination clinics to administer vaccine to as many target subjects as possible in a time-efficient way.
• Promote vaccination of family members of at-risk patients and health-care personnel.
• Display appropriate information in your patients' waiting room and in your office.

Table 19. Possible actions by primary-care physicians to encourage vaccine uptake in target populations. Courtesy of Ted van Essen on behalf of the Dutch College of General Practitioners.

threat. At the same time, influenza vaccination is frequently considered as ineffective or even a cause of the flu. In addition, the current mode of influenza vaccine administration by injection represents a barrier for individuals with fear of needles.

To improve the vaccination coverage rates in target groups, in accordance with WHO recommendations,[10,11,36] it is important that existing national vaccination policies are effectively implemented. National health authorities of eight different European countries have recently discussed possible ways to reach the WHO vaccine coverage rate objectives for 2010.[29] Health-care professionals play the single most important role in making this happen. Indeed, they are in the best position to educate and motivate patients to be vaccinated. Table 19 lists a number of recommendations in this respect. As indicated in Chapter 3 and further described in Chapter 9, there is a societal urgency to achieve the WHO objectives,[36] because this will not only ameliorate the annual burden of influenza but also contribute to a better level of pandemic preparedness.

Vaccination of healthy younger adults

While younger adults are generally not at risk for serious complications due to influenza, flu remains an important cause of work absenteeism, diminished work productivity and malaise interfering with off-work activities. This is why there is an increasing awareness of the potential benefits of vaccination of working adults.[12,13] Several prospective clinical studies have demonstrated the efficacy of inactivated influenza vaccines among healthy younger adults. Initial trials, conducted among military recruits several decades ago,[57] showed that the vaccine was 70–90% efficacious in preventing laboratory-confirmed influenza,

Benefits of influenza vaccination of healthy adults and children	
Outcome measure	**Reduction (%)**
Healthy adults <65 years of age	
Laboratory-confirmed influenza	70–90
URI/ILI (all causes)	25–34
Work loss due to URI/ILI	32–43
Physician visits due to URI/ILI	42–44
Children	
Laboratory-confirmed influenza	60–90
Acute otitis media (all causes)	30–36

Table 20. Benefits of influenza vaccination of healthy adults and children. ILI, influenza-like illness; URI, upper respiratory illness. Adapted from Nichol KL. The efficacy, effectiveness and cost-effectiveness of inactivated influenza virus vaccines. *Vaccine* 2003; **21**: 1769–1775[12] with permission from Elsevier.

provided there was a good antigenic match between vaccine and circulating virus.[58] A review of more recent clinical studies shows that the efficacy of inactivated influenza vaccines varied from 65% for all influenza seasons to 72% for those seasons where there was a good match between vaccine and circulating virus.[31] Additional studies have reported vaccine efficacies in terms of prevention of confirmed influenza in the range of 80–90% in cases where there was a good match.[59,60] Clearly, current inactivated influenza vaccines attain very high efficacy values among healthy younger adults (Table 20).

As demonstrated by a number of studies, conducted in different countries, vaccination significantly reduces illness, absenteeism and influenza-related costs for healthy adults in the work place.[12,13] Indeed, vaccination reduces upper respiratory tract and influenza-like illnesses from all causes by approximately 30%, related physician visits by >40% and work loss by >35% (Table 20).[60,61] Accordingly, cost–benefit analyses, based on clinical trials or on modelling, have shown that vaccination of healthy working adults is cost-effective and in many cases cost-saving, provided that indirect costs associated with work absenteeism (see Chapter 6) are explicitly taken into account.[62] For example, trials conducted in the USA have shown that – with an average cost for vaccine production and administration of $20 – the net saving would be $23 per person vaccinated.[63] In another study comparing 131 vaccinated employees from six textile plants in North Carolina, USA, with 131 age- and gender-matched non-vaccinated controls from different plants, the "cost per saved lost work day" was $22.36, resulting in an overall saving of $2.58 per dollar invested in the vaccination programme.[64] Other, model-based, studies also indicate that vaccinating working adults would be cost-saving.[12] While recent international guidelines for pharmacoeconomic analyses do explicitly recommend the inclusion of production gains and losses,[62] also when such indirect costs are not taken into account, vaccination of adults below the age of 65 turns out to be highly cost-effective. For example, a cost-utility analysis of the US Office of Technology Assessment found in 1981 that a year of healthy life gained would cost $278 per person of 25–44 years of age and only $100 for those 45–64 years old.[65] Recent studies in Europe have also shown the cost-effectiveness of vaccinating everyone over 50 years of age compared to current age recommendations in most countries.[30,31]

Vaccination of children

In most countries, healthy toddlers and children are not included in the target groups for annual influenza immunization. However, there is increasing epidemiological evidence of the burden of disease in children[33] and of vaccination effectiveness.[33,35,66] In addition to the direct benefits for the vaccinated children, a vaccination programme for children may also have the potential for reducing the impact of influenza epidemics, because children play an important role in the spread of influenza infections in communities.[66,67] In addition, influenza among children is a significant cause of parental work loss. Furthermore, very small children may well be at increased risk for serious influenza-associated complications. For this reason, the Advisory Committee on Immunization Practices (ACIP) in the USA now recommends routine vaccination of children of 6–23 months of age.[41] A similar recommendation has been formulated recently in a consensus report by a European vaccination expert group.[33]

The efficacy of influenza vaccination among children has been evaluated in a number of randomized, controlled trials, involving the use of either trivalent inactivated or experimental live-attenuated vaccines.[68–70] From these studies, it appears that vaccination is highly efficacious in terms of preventing laboratory-confirmed influenza for children in their teens (~90%), whereas a lower efficacy is seen with younger children (Table 20). For example, an Italian study among children 1–6 years of age showed a reduction by 67% in ILI. Other studies generally confirm this picture.[69] The comparatively modest immunogenicity of influenza vaccines among small children is probably due to their lack of pre-exposure to either influenza virus or vaccine. Therefore, for these immunologically naive children, usually a two-dose vaccination regimen is recommended.[41]

A common complication of influenza among young children is acute otitis media (see Chapter 5). There appears to be a clear-cut benefit associated with influenza vaccination in terms of reducing the incidence of otitis media. A clinical study conducted in Finland, for example, has demonstrated a reduction of 36% in overall rate of otitis media as a result of vaccination (Table 20), which corresponded to an 83% reduction among children with laboratory-confirmed influenza.[71]

Another benefit of vaccinating children relates to the associated prevention of secondary transmission among family members or others within the neighbourhood. In a study conducted in Michigan, USA, during the 1968–69 pandemic outbreak of Hong Kong flu, vaccination of school-age children resulted in three-fold lower rates of influenza-like illness than in a control community.[72] Interestingly, a 20-year programme in Japan, involving vaccination of school-age children, has indicated that there may be a correlation between increased vaccination of children and lower excess mortality among the elderly,[67] substantiating the notion that vaccination of children reduces secondary influenza transmission. A similar observation has been made in a recent study in Russia.[35] School children in some regions in the Moscow area were systematically vaccinated with a classical inactivated subunit vaccine, whereas in control areas no such vaccination strategy was installed. This 2-year study also clearly demonstrated the benefits of vaccination of children.[35]

Finally, vaccination of children appears to be highly cost-effective and in many cases cost-saving. The analysis by the US Office of Technology Assessment, referred to above,[65] indicates that vaccination of children below 3 years of age would cost US $1122 per year of healthy life gained, while vaccinating children 3–14 years of age would only cost US $853 per year of healthy life gained. More recent analyses also clearly indicate the economic benefits of vaccination of children against influenza.[73] Even at an average cost of influenza vaccine production and administration of US $20, vaccination of children is likely to be cost-saving, part of the benefit in this respect being due to a reduction in parental work loss.[12]

Towards universal flu vaccination

As we have described in this chapter and elsewhere in the book, influenza is a major cause of illness and suffering, not only for the elderly and people with underlying medical conditions, but also for healthy younger adults and children. While younger adults are generally not at risk for serious complications due to influenza, flu remains an important cause of work absenteeism, diminished work productivity and malaise interfering with non-work activities. The economic burden of disease in healthy adults, as well as the demonstrated medical and economical benefits of vaccination, justify including this segment of the population in the recommended target groups for routine immunization.

Accumulating epidemiological data of the burden of disease in young children, together with the demonstrated direct and indirect benefits of vaccination programmes for very young and school-aged children, would justify also including this segment of the population in the recommendations for routine immunization. Based on such considerations, the European Scientific Working group on Influenza (ESWI) has recently estimated that at least 30% of the total population should be annually vaccinated against influenza.[74]

References

1. Smith W, Andrews CH, Laidlaw PP. A virus obtained from influenza patients. *Lancet* 1933; **ii**: 66–68.

2. Wood JM, Williams MS. History of inactivated influenza vaccines. In: Nicholson KG, Webster RG, Hay AJ, editors. *Textbook of Influenza*. Blackwell Science, 1998; pp. 317–323.

3. Francis T, Magill TP. The incidence of neutralizing antibodies for human influenza virus in the serum of human individuals of different ages. *J Exp Med* 1936; **63**: 655–668.

4. Furminger IGS. Vaccine production. In: Nicholson KG, Webster RG, Hay AJ, editors. *Textbook of Influenza*. Blackwell Science, 1998; pp. 324–332.

5. Wood JM. Standardization of inactivated influenza vaccines. In: Nicholson KG, Webster RG, Hay AJ, editors. *Textbook of Influenza*. Blackwell Science, 1998; pp. 333–345.

6. Hobson D, Curry RL, Beare AS, Ward-Gardner A. The role of serum haemagglutination-inhibiting antibody in protection against challenge virus infection with A2 and B viruses. *J Hyg (Lond)* 1972; **70**: 767–777.

7. Small PR, Waldman RA, Bruono JC, Gifford GE. Influenza infection in ferrets: Role of serum antibody in protection and recovery. *Infect Immun* 1976; **13**: 417–424.

8. Ghendon,Y. Influenza surveillance. *Bull World Health Org* 1991; **61**: 509–515.

9. World Health Organization. Influenza vaccines. *Wkly Epidemiol Rec* 2000; **75**: 281–288.

10. The Macroepidemiology of Influenza Vaccination (MIV) Study Group. The macroepidemiology of influenza vaccination in 56 countries, 1997–2003. *Vaccine* 2005; **23**: 5133–5143.

11. World Health Organization. Influenza vaccines. *Wkly Epidemiol Rec* 2002; **77**: 230–239.

12. Nichol KL. The efficacy, effectiveness and cost-effectiveness of inactivated influenza virus vaccines. *Vaccine* 2003; **21**: 1769–1775.

13. Nichol KL. Efficacy/clinical effectiveness of inactivated influenza virus vaccines in adults. In: Nicholson KG, Webster RG, Hay AJ, editors. *Textbook of Influenza*. Blackwell Science, 1998; pp. 358–372.

14. Wiselka MJ. Vaccine safety. In: Nicholson KG, Webster RG, Hay AJ, editors. *Textbook of Influenza*. Blackwell Science, 1998; pp. 346–357.

15. Stöhr K. The global agenda on influenza surveillance and control. *Vaccine* 2003; **21**: 1744–1748.

16. World Health Organization. Global agenda on influenza – adopted version. Part I, *Wkly Epidemiol Rec* 2002; **77**: 179–182; Part II, *Wkly Epidemiol Rec* 2002; **77**: 191–196. (www.who.int/emc/diseases/flu/global_agenda_report/Contentpandemic.htm)

17. Ruben FL. Prevention and control of influenza. Role of vaccine. *Am J Med* 1987; **82**(Suppl 6a): 31–34.

18. Nicholson KG, Tyrrell DAJ, Harrison P *et al.* Clinical studies of monovalent inactivated whole virus and subunit A/USSR/77 (H1N1) vaccine: serological responses and clinical reactions. *J Biol Stand* 1979; **7**: 123–136.

19. Beyer WEP, Palache AM, Osterhaus ADME. Comparison of serology and reactogenicity between influenza subunit vaccines and whole or split vaccines. A review and meta-analysis of the literature. *Clin Drug Invest* 1998; **15**: 1–12.

20. Smith DJ, Lapedes AS, de Jong JC *et al.* Mapping the antigenic and genetic evolution of influenza virus. *Science* 2004; **305**: 371–376.

21. Palache AM, Beyer WEP, Lüchters G *et al.* Influenza vaccines: the effect of vaccine dose on antibody response in primed populations during the ongoing interpandemic period. A review of the literature. *Vaccine* 1993; **11**: 892–908.

22. Treanor J, Keitel W, Belshe R *et al.* Evaluation of a single dose of half strength inactivated influenza vaccine in healthy adults. *Vaccine* 2002; **20**: 1099–1105.

23. European Medicines Agency (EMEA). Note for guidance on harmonization of requirements for influenza vaccines.

CPMP/BWP/214/96, 1997. (www.emea.eu.int/pdfs/human/bwp/021496en.pdf)

24. Wood JM, Levandowski RA. The influenza vaccine licensing process. *Vaccine* 2003; **21**: 1786–1788.

25. Voordouw B. *Influenza Vaccination in Community Dwelling Elderly Persons*. PhD thesis, Erasmus University, Rotterdam, 2005. ISBN 90-8559-103-1.

26. Gerdil C. The annual production cycle for influenza vaccine. *Vaccine* 2003; **21**: 1776–1779.

27. Kroneman MW, van Essen GA, Tacken MA *et al.* Does a population survey provide reliable influenza vaccine uptake rates among high-risk groups? A case-study of The Netherlands. *Vaccine* 2004; **22**: 2163–2170.

28. Kroneman M, van Essen GA, John Paget W. Influenza vaccination coverage and reasons to refrain among high-risk persons in four European countries. *Vaccine* 2006; **24**: 622–628.

29. Szucs TD, Muller D. Influenza vaccination coverage rates in five European countries – a population-based cross-sectional analysis of two consecutive influenza seasons. *Vaccine* 2005; **23**: 5055–5063.

30. Chancellor J. Lowering the age threshold for routine influenza vaccination to 50 years: cost-effectiveness analysis for European countries. *Vaccine* 2006; in press.

31. Turner DA, Wailoo AJ, Cooper NJ *et al.* The cost-effectiveness of influenza vaccination of healthy adults 50–64 years of age. *Vaccine* 2006; **24**: 1035–1043.

32. Centers for Disease Control and Prevention. Influenza vaccination in pregnancy: practices among obstetrician-gynecologists – United States, 2003–04 influenza season. *MMWR* 2005; **54**: 1050–1052. (www.cdc.gov/mmwr/preview/mmwrhtml/mm5441a4.htm)

33. Heikkinen T, Booy R, Campins M *et al.* Should healthy children be vaccinated against influenza? A consensus report of the Summits of Independent European Vaccination Experts. *Eur J Pediatr* 2005; **21**: 1–6.

34. Heikkinen T, Ruuskanen O. Influenza vaccines in healthy children (letter to the editor). *Lancet* 2005; **365**: 2086–2087.

35. Ghendon YZ, Kaira AN, Elshina GA. The effect of mass influenza immunization in children on the morbidity of the unvaccinated elderly. *Epidemiol Infect* 2006; **134**: 71–78.

36. Resolution WHA56.19. Prevention and control of influenza pandemics and annual epidemics. Fifty-Sixth World Health Assembly, Geneva, 19–28 May 2003. (www.who.int/gb/ebwha/pdf_files/WHA56/ea56r19.pdf)

37. Smith SA, Poland GA. The use of influenza and pneumococcal vaccines in people with diabetes (technical review). *Diabetes Care* 2000; **23**: 95–108.

38. Centers for Disease Control and Prevention. Influenza vaccination levels among person aged ≥65 years and among persons aged 18–64 years with high-risk conditions – United States, 2003. *MMWR* 2005; **54**: 1045–1049. (www.cdc.gov/mmwr/preview/mmwrhtml/mm5441a3.htm)

39. Fedson DS, Nichol K. Should we question the benefits of influenza vaccination for elderly people? *Infectious Disease News*, Guest Editorial, August 2005. (www.infectiousdiseasenews.com/200508/guested1.asp)

40. Voordouw AC, Sturkenboom MC, Dieleman JP *et al.* Annual revaccination against influenza and mortality risk in community-dwelling elderly persons. *J Am Med Assoc* 2004; **292**: 2089–2095.

41 Harper SA, Fukuda K, Uyeki TM *et al.* Advisory Committee on Immunization Practices (ACIP), Centers for Disease Control and Prevention (CDC). Prevention and control of influenza. Recommendations of the Advisory Committee on Immunization Practices (ACIP). *MMWR Recomm Rep* 2005; **54**(RR-8): 1–40.

42. Nichol KL, Nordin J, Mullooly J *et al.* Influenza vaccination and reduction in hospitalizations for cardiac disease and stroke among the elderly. *New Engl J Med* 2003; **348**: 1322–1332.

43. Nichol KL, Mendelman P. Influence of clinical case definitions with differing levels of sensitivity and specificity on

estimates of the relative and absolute health benefits of influenza vaccination among healthy working adults and implications for economic analyses. *Virus Res* 2004; **103**: 3–8.

44. Beyer WEP. Heterogeneity of case-definitions used in vaccine effectiveness studies and its impact on meta-analysis. *Vaccine* 2006; in press.

45. Demicheli V, Rivetti D, Deeks JJ *et al.* Vaccines for preventing influenza in healthy adults. *Cochrane Database Syst Rev* 2004; **3**: CD001269 Review.

46. Jefferson T, Smith S, Demicheli V *et al.* Assessment of the efficacy and effectiveness of influenza vaccines in healthy children: systematic review. *Lancet* 2005; **365**: 773–780.

47. Nichol K, Nordin J, Mullooly J. Influence of clinical outcome and outcome period definitions on estimates of absolute clinical and economic benefits of influenza vaccination in community dwelling elderly persons. *Vaccine* 2006; **24**:1562–1568

48. Govaert TME, Thijs CTMCN, Masurel N *et al.* The efficacy of influenza vaccination in elderly individuals. A randomized double-blind placebo-controlled trial. *J Am Med Assoc* 1994; **272**: 1661–1665.

49. Nichol KL. Influenza vaccination in the elderly. Impact on hospitalization and mortality. *Drugs Aging* 2005; **22**: 495–515.

50. Vu T, Farish S, Jenkins M, Kelly H. A meta-analysis of effectiveness of influenza vaccine in persons aged 65 years and over living in the community. *Vaccine* 2002; **20**: 1831–1836.

51. Gross PA, Hermogenes AW, Sachs HS, Lau J, Levandowski RA. The efficacy of influenza vaccine in elderly persons: a meta-analysis and review of the literature. *Ann Intern Med* 1995; **123**: 518–527.

52. Postma MJ, Bos JM, Van Gennip M *et al.* Economic evaluation of influenza vaccination. Assessment for The Netherlands. *Pharmacoeconomics* 1999; **16**(Suppl 1): 33–40.

53. Watson JM, Cordier JF, Nicholson KG. Does influenza immunisation cause exacerbations of chronic airflow obstruction or asthma? *Thorax* 1997; **52**: 190–194.

54. Park CL, Frank AL, Sullivan M *et al.* Influenza vaccination of children during acute asthma exacerbation and concurrent prednisone therapy. *Pediatrics* 1996; **98**: 196–200.

55. Safranek TJ, Lawrence DN, Kurland LT *et al.* Reassessment of the association between Guillain-Barré syndrome and receipt of swine influenza vaccine in 1976–1977: results of a two-state study. *Am J Epidemiol* 1991; **133**: 940–951.

56. Lasky T, Terracciano GJ, Magder L *et al.* The Guillain-Barré syndrome and the 1992–1993 and 1993–1994 influenza vaccines. *New Engl J Med* 1998; **339**: 1797–1802.

57. Meiklejohn G, Kempe CH, Thalman WG *et al.* Effectiveness of polyvalent influenza A vaccine during an influenza A-prime epidemic. *Am J Hyg* 1954; **59**(3): 241–248.

58. Demicheli V, Jefferson T, Rivetti D *et al.* Prevention and early treatment of influenza in healthy adults. *Vaccine* 2000; **18**: 957–1030.

59. Wilde JA, McMillan JA, Serwint J *et al.* Effectiveness of influenza vaccine in health care professionals: a randomized trial. *J Am Med Assoc* 1999; **281**: 908–913.

60. Bridges CB, Thompson WW, Meltzer MI *et al.* Effectiveness and cost-benefit of influenza vaccination of healthy working adults: a randomized controlled trial. *J Am Med Assoc* 2000; **284**: 1655–1663.

61. Nichol KL, Lind A, Margolis KL *et al.* The effectiveness of vaccination against influenza in healthy, working adults. *New Engl J Med* 1995; **333**: 889–893.

62. Postma MJ, Jansema P, van Genugten ML *et al.* Pharmacoeconomics of vaccinating healthy working adults against influenza; reviewing the available evidence. *Drugs* 2002; **62**: 1013–1024.

63. Nichol KL, Mallon KP, Mendelman PM. Cost benefit of influenza vaccination in healthy, working adults: an economic analysis based on the results of a clinical trial of trivalent live attenuated influenza virus vaccine. *Vaccine* 2003; **21**: 2207–2217.

64. Campbell DS, Rumley MH. Cost-effectiveness of the influenza vaccine in a healthy, working-age population. *J Occup Environ Med* 1997; **5**: 408–414.

65. Office of Technology Assessment. Cost-effectiveness of influenza vaccination. Washington, DC: Congress of the United States, 1981.

66. Jordan R, Connock M, Albon E *et al.* Universal vaccination of children against influenza: Are there indirect benefits to the community? A systematic review of the evidence. *Vaccine* 2006; **24**: 1047–1062.

67. Reichert TA, Sugaya N, Fedson DS *et al.* The Japanese experience with vaccinating schoolchildren against influenza. *New Engl J Med* 2001; **344**: 889–896.

68. Neuzil KM, Dupont WD, Wright PF, Edwards KM. Efficacy of inactivated and cold-adapted vaccines against influenza A infection, 1985 to 1990: the pediatric experience. *Pediatr Infect Dis J* 2001; **20**: 733–740.

69. Neuzil KM, Edwards KM. Influenza vaccines in children. *Semin Pediatr Infect Dis* 2002; **13**: 174–181.

70. Schonbeck Y, Sanders EA, Hoes AW *et al.* Rationale and design of the prevention of respiratory infections and management in children (PRIMAKid) study: a randomized controlled trial on the effectiveness and costs of combined influenza and pneumococcal vaccination in pre-school children with recurrent respiratory tract infections. *Vaccine* 2005; **23**: 4906–4914.

71. Heikkinen T, Ruuskanen O, Waris M *et al.* Influenza vaccination in the prevention of acute otitis media in children. *Am J Dis Child* 1991; **145**: 445–448.

72. Monto AS, Davenport FM, Napier JA, Francis T. Modification of an outbreak of influenza in Tecumseh, Michigan by vaccination of schoolchildren. *J Infect Dis* 1970; **122**: 16–25.

73. Cohen GM, Nettleman MD. Economic impact of influenza vaccination in preschool children. *Pediatrics* 2000; **106**: 973–976.

74. European Scientific Working Group on Influenza (ESWI). Influenza vaccination for one third of the population of the European Union (EU) 25 Member States by 2010, 2005. (www.eswi.org)

9. Vaccination: Novel Developments and Pandemic Vaccines

Current inactivated influenza vaccines significantly reduce influenza morbidity and mortality among at-risk populations, as discussed in the previous chapter. Furthermore, these vaccines have an excellent safety record. However, limitations remain. Firstly, the efficacy of inactivated vaccines, particularly in the elderly, is lower than that of many other prophylactic viral vaccines. Secondly, current vaccine formulations are not efficacious enough to induce adequate protection against entirely new pandemic virus subtypes. Thirdly, the current route of vaccine administration by intramuscular injection may be a barrier for acceptance of the vaccine, especially with children. Finally, vaccine virus growth in embryonated chicken eggs poses limitations to the flexibility of the vaccine production process and to the production capacity, particularly under conditions of a sudden high demand for vaccine, such as in the case of an emerging pandemic.

It is clear, therefore, that there is a need for novel, more efficacious, influenza vaccine modalities.[1–5] These will improve the clinical effectiveness of annual flu vaccination in current target groups, who stand to derive great benefit from more robust immunological responses to better vaccines. More efficacious vaccine formulations will also contribute significantly to our pandemic preparedness. Vaccine formulations must be developed that may be delivered via routes other then injection – for example, locally with a nasal spray. Various novel influenza vaccine modalities, registered and experimental, as well as new approaches for vaccine virus production, will be presented below. Clearly, while ideas and concepts have been manifold, only a few novel vaccines have made it to the

Key Messages

- Increase of annual influenza vaccine use according to WHO recommendations is necessary to optimize the benefits of vaccination for individuals and society as well as to increase the global level of pandemic preparedness.
- While current inactivated influenza vaccines substantially reduce influenza morbidity and mortality, there remains a need for novel influenza vaccines with improved efficacy, particularly among the elderly.
- Novel vaccine modalities for the elderly should induce not only a robust antibody response, but also T cell-mediated immunity.
- Since injection represents a barrier for optimal acceptance of influenza vaccine, there is a need for vaccine formulations that may be administered by different routes.
- Vaccines administered intranasally would have the additional advantage of inducing local immunity in the respiratory tract.
- Growth of vaccine virus on cell-culture substrates will considerably improve vaccine supply capabilities, which is of particular importance under conditions of pandemic threat.
- Equitable accessibility of pandemic vaccines is best guaranteed if recommendations for annual influenza vaccine usage are fully implemented and low-dose adjuvanted pandemic vaccines are developed and registered.

marketplace in recent years, including a live-attenuated virus vaccine for intranasal delivery and a limited number of injectable vaccine presentation forms or adjuvanted vaccines. This underscores the fact that the path from conception to registration is often long and full of hurdles, many new vaccine modalities failing along the way. It also underscores the need to take the challenge posed by pandemic vaccines seriously. Their development is a long-term activity and failures are to be expected, while the next pandemic appears imminent (see Chapter 3).

Novel inactivated influenza vaccines with improved efficacy

One of the limitations of current inactivated influenza vaccines is their suboptimal efficacy among the elderly,[6–8] which appears to be primarily related to a diminishing T cell activity with age (see Chapter 4).[9] These age-related changes in cell-mediated immunity put specific demands on novel vaccine modalities. The presentation form of conventional vaccines is not optimal for stimulation of cell-mediated immunity, particularly cytotoxic T lymphocyte (CTL) activity. Moreover, as opposed to the viral HA that directs the antibody response against influenza, the antigenic determinants of CTL activity are primarily derived from the internal proteins of the virus (see Chapter 4).

It is for these reasons that, particularly in the case of new vaccines for the elderly, improvements of the formulations are sought at the level of T cell immunity, in addition to induction of antibody responses. Accordingly, antibody titres are unlikely to remain the sole reliable correlate of protection.[10] In evaluating the efficacy of novel generations of flu vaccines, it will therefore be necessary to consider, besides antibody titres, also other parameters, including T cell-mediated immunity.[11]

Current inactivated vaccines exhibit relatively poor immunogenicity in immunologically naive individuals, such as small children, who have not had prior exposure to virus or vaccine.[12] Likewise, a single dose of current, non-adjuvanted vaccine formulation provides inadequate protection in case of a pandemic,[13–17] again as a result of a total lack of pre-exposure of the population to the emerging new virus subtype. More powerful vaccines are needed which, in these cases, primarily need to induce a robust antibody response against HA.[17]

Adjuvanted vaccines

The use of adjuvants represents one way to improve the efficacy of vaccines. An adjuvanted influenza vaccine, Fluad®, has been licensed in Europe in recent years. It is formulated as an oil-in-water emulsion, referred to as MF59, containing squalene, Tween 80 and sorbitan trioleate. This vaccine, which contains the usual 15 μg of HA per strain, has a modest immunopotentiating effect in primed individuals, including the elderly.[18] On the other hand, the MF59-adjuvanted vaccine appears to be more reactogenic than non-adjuvanted vaccine.[18]

A promising class of adjuvants is derived from bacterial enterotoxins, such as cholera toxin (CT) and the *E. coli* heat-labile enterotoxin (LT).[19] These molecules, including a number of detoxified CT or LT variants, are of particular interest in the context of mucosal vaccine delivery and potentiation of local secretory IgA responses, as discussed below. However, the non-toxic recombinant B subunit of LT (LTB) also has potent adjuvant activity when administered intramuscularly to mice along with influenza subunit vaccine.[20]

Immune stimulation can also be induced by the use of specific cytokines, including IL-2 and GM-CSF. In murine preclinical studies, these cytokine adjuvants strongly promoted antibody responses to influenza surface glycoproteins, including neuraminidase, in a liposome formulation, mediating cross-reactive protection against a variety of influenza A strains.[21] A similar experimental vaccine has recently been evaluated in human clinical trials.[22] Stringent safety testing seems mandatory, since cytokines may well have multiple biological effects besides adjuvant activity.

Virosomes as a novel vaccine presentation system

The immune response to inactivated viral vaccines can also be improved by modification of the way in which the antigen is presented to the immune system. Virosomes are a presentation form that closely mimics the native virus. Virosomes are virus-like particles lacking the viral genetic material. They are generated by a reconstitution procedure,[23] involving treatment of influenza virus with an appropriate detergent such that the viral envelope is selectively solubilized. After separation of the nucleocapsid fraction by ultracentrifugation, the detergent is removed from the

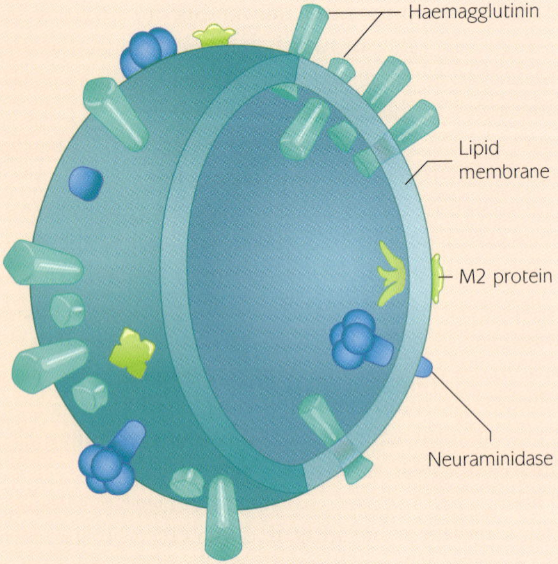

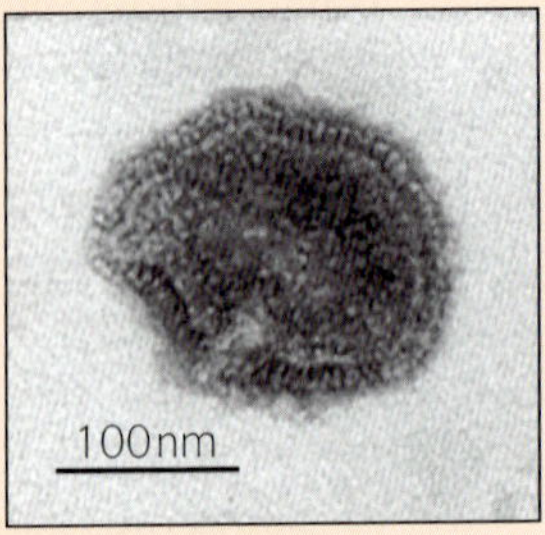

Figure 29. (Top) Model of a virosome derived from influenza virus. (Bottom) Electron micrograph of a virosome. Image courtesy of Laura Bungener, University Medical Centre, Groningen, the Netherlands.

supernatant in a controlled fashion such that viral envelopes ("virosomes") are reformed. Figure 29 shows a schematic representation and an electron micrograph of influenza virosomes.

Functionally reconstituted virosomes preserve the receptor-binding and membrane fusion activity of the viral HA. This implies that these virosomes, like the native virus, are internalized by cells via receptor-mediated endocytosis and fusion from within acidic endosomes (see Chapter 2). Through this fusion process, a continuity is established between the virosomal lumen and the cell cytosol, while the HA is introduced into the endosomal membrane.

During the preparation of virosomes, biologically active molecules, including protein antigens, may be encapsulated in the virosomal lumen. Fusion of the virosomes from within the endosomal cell compartment will result in delivery of these virosome-encapsulated compounds to the target cell cytosol. Accordingly, interaction of virosomes with antigen-presenting cells (APCs), such as dendritic cells (DCs), results in delivery of virosome-encapsulated protein antigen to the lumen of the cells. In this manner, the antigen gains access to the MHC class I presentation pathway[24] (Figure 30). In addition, interaction of virosomes with DCs results in activation and maturation of DCs.[25] *In vivo* studies in mice have demonstrated that fusion-active virosomes, containing an encapsulated protein antigen, efficiently prime a class I MHC-restricted CTL response against the encapsulated antigen.[25] In addition, virosomes induce a potent antibody response against the virosomal HA. Thus, virosomes induce a broad immune response, priming both the humoral and cellular arms of the immune system.[26,27]

Currently, there are two virosome-based influenza vaccines, Inflexal V® and Invivac®, on the market. These vaccines have excellent safety and immunogenicity profiles.[28,29] In recently conducted comparative trials, the MF59-adjuvanted vaccine Fluad® and the two virosomal vaccine formulations showed similar efficacy and immunogenicity.[30,31] As observed earlier,[18] the MF59-adjuvanted vaccine was somewhat more reactogenic than the virosomal vaccines.

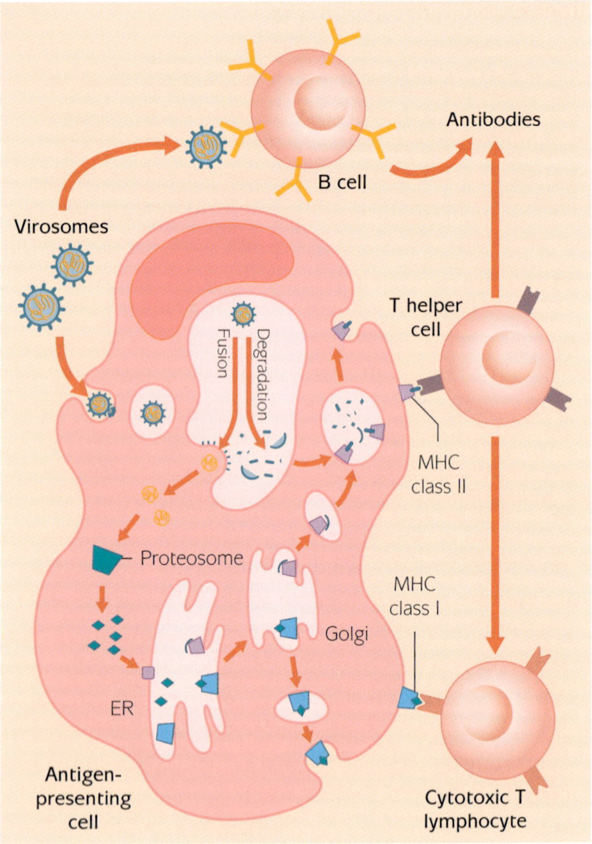

Figure 30. Interaction of virosomes with cells of the immune system. By virtue of the repetititve arrangement of HA on the virosome surface, virosomes interact efficiently with immunoglobulin receptors on B lymphocytes. Virosomes are also taken up avidly by antigen-presenting cells, particularly dendritic cells. Antigens on the virosome surface, as well as antigens derived from degraded virosomes, enter the MHC class II pathway, activating T helper cells. Antigens inside the virosomes, through fusion of the virosomes, access the cytosolic MHC class I presentation pathway, activating cytotoxic T lymphocytes. Adapted from Huckriede A *et al*. Influenza virosomes: combining optimal presentation of hemagglutinin with immunopotentiating activity. *Vaccine* 2003; **21**: 925–931 with permission from Elsevier.

Other experimental approaches

Like virosomes, immune-stimulating complexes (ISCOMs) also represent a presentation form for membrane-associated antigens. ISCOMs are cage-like structures formed from phospholipids, cholesterol and saponins, the latter functioning not only as an essential component of the ISCOM matrix, but also as a built-in adjuvant. Integral membrane proteins, such as influenza HA, may be incorporated in the ISCOM matrix during their preparation. ISCOMs have been shown to strongly promote antibody responses to incorporated protein antigens in a number of preclinical animal models, including a non-human primate model.[32] Also, ISCOMs have the capacity to introduce antigens into the MHC class I presentation pathway, thus inducing CD8-positive CTL activity.[33] An ISCOM-based influenza vaccine for horses is on the market. In a human influenza vaccine trial, however, ISCOMs did not show enhanced antibody responses as compared to conventional inactivated influenza vaccines.[34] In addition, there remains an issue with regard to the reactogenicity of the vaccine due to the presence of the saponin component in the ISCOM structure.

Several other experimental approaches for the development of novel, more effective, influenza vaccine modalities are under investigation. These include DNA immunization,[35,36] and experimental vaccines targeted at conserved viral antigens, such as M2 and/or NP, with the purpose of inducing an immune response that would be cross-reactive to different influenza drift variants or even subtypes. DNA immunization relies on the *in situ* synthesis of protein antigens after intramuscular administration of naked DNA. As protein antigens may also be synthesized within APCs, DNA immunization is expected to access the MHC class I presentation pathway and thus to stimulate CTL activity. DNA immunization against a number of influenza antigens, including HA and NP, has given promising results in murine model systems, but appears less effective in humans. In addition, safety concerns remain with regard to the use of DNA for mass vaccination purposes.

Promising results have been obtained with the use of an experimental M2-based vaccine in mice.[37] However, a more recent study in pigs has raised considerable concern, in that animals vaccinated with an experimental M2/NP DNA vaccine exhibited exacerbation of disease upon challenge, rather than protection.[38] Clearly, more data are needed to determine whether M2 represents an appropriate antigen for use in novel flu vaccine modalities.

Mucosal delivery of influenza vaccines

To those of us fearing needles, the current mode of vaccine administration by injection may represent a barrier for taking an annual flu shot. This holds true in particular for small children, a group under consideration for routine annual flu vaccination (see Chapter 8). In addition, administration of the vaccine by intramuscular injection is not in line with the site of entry of the influenza virus via mucosal surfaces of the nasal cavity, and does not generally induce a local mucosal secretory IgA (sIgA) response as the first line of defence against incoming virus (see Chapter 4). It is for these reasons that there is considerable interest in influenza vaccines for intranasal delivery. Figure 31 shows a simple spray device for administration of such vaccines.

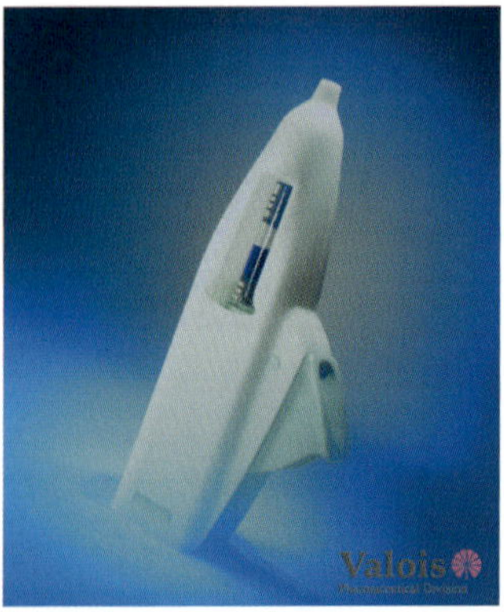

Figure 31. Spray device for intranasal administration of influenza vaccine. Courtesy of Valois Pharmaceutical Division, Marly-le Roi, France.

A major hurdle in the development of intranasal vaccine formulations has been the comparatively low immunogenicity of mucosal vaccines,[39] especially when induction of serum HI antibodies is taken as the primary correlate of protection. This problem can be overcome by the use of live attenuated influenza virus preparations, such as cold-adapted viruses that replicate in the upper respiratory tract, inducing both local and systemic antibody responses as well as cellular immunity. Indeed, as discussed below, a live attenuated influenza vaccine (LAIV) is currently registered and on the market in the USA under the name of FluMist®. In addition, there are several candidate inactivated influenza vaccines for intranasal delivery under development. Because of the low immunogenicity of mucosally delivered vaccines,[39] such inactivated formulations require the use of an adjuvant.

Live attenuated influenza vaccine

Intranasally delivered LAIV closely mimics a natural infection. LAIVs have been used for decades in Russia,[40] but unfortunately not much information about the safety and efficacy of these formulations is available. In the 1960s, a research programme was initiated in the USA to develop LAIVs. In 2003, this has led to the approval of FluMist® in the USA, for healthy individuals between ages 5 and 49 years.

LAIVs are made by reassortment of a temperature-sensitive, cold-adapted (*ca*), parent virus with the actual vaccine virus strain. The *ca* virus replicates at the reduced temperatures of the upper respiratory tract, but in the lungs at 37°C replication is inhibited. As a result, the virus has lost its virulence in the human host and does not cause clinical influenza. Because of the intrinsic unpredictability of the formation of natural reassortants, extensive safety research has been done to demonstrate that the *ca* viruses are genetically stable and do not convert back to the wild-type virus.[41] Based on these studies, the chances of a sudden emergence of a pathogenic influenza virus following the use of the live virus vaccine are considered remote.

Many placebo-controlled studies with LAIV have been carried out, among children as well as adults.[41–43] These studies have shown that the LAIV is safe and efficacious. Some studies suggest that the LAIV might induce a certain degree of cross-protection against drift viruses, possibly due to its ability to induce a CTL response.[43] Another potential advantage of the LAIV may be its suggested ability to dampen the spread of virus in the community, particularly if the vaccine were to be used widely among children.[41] However, theoretically, the same would be true for conventional inactivated vaccines. Clearly, the potential advantages of LAIVs over inactivated vaccines need further scientific demonstration and confirmation.

In several clinical trials, intranasal LAIV has been compared directly with conventional intramuscularly administered inactivated vaccines. A meta-analysis of these vaccine-controlled clinical trials has shown that mucosal sIgA antibody titres were higher after vaccination with LAIV, while inactivated vaccines induced higher serum IgG antibody titres.[44] In studies where culture-positive influenza illness was assessed, both vaccine types were found to be equally effective.

Since the introduction of FluMist® in the USA in 2003, there has been concern about the use of live-virus vaccines in immunocompromised patients. Indeed, health-care workers vaccinated with LAIV have been advised not to be in contact with such patients for up to 3 weeks.[45] Clearly, more experience with this vaccine is required to relax these concerns. If the theoretical advantages of the LAIV can be scientifically substantiated and the safety concerns about the use of LAIV prove unwarranted,[41] FluMist® may well offer interesting new possibilities to control annual influenza epidemics. As the frozen formulation of FluMist® has been an obstacle for efficient distribution of the vaccine, a second-generation refrigerator-stable version of the vaccine is being developed; this formulation has recently been evaluated in clinical trials among children between 6 and 59 months of age.

Besides FluMist®, there are other LAIV modalities, based on cold-adapted viruses, under development.[46]

In addition to the cold-adapted virus vaccines, novel recombinant influenza viruses, designed by modern reverse genetics techniques (see below), have been advocated as potential LAIVs. These include viruses with deletions in the NS1, enhancing their interferon sensitivity,[5] replication-incompetent NS2 knockouts[47] or viruses with attenuating mutations in M2.[48]

Candidate inactivated intranasal vaccines

A number of candidate adjuvanted inactivated influenza vaccines for intranasal administration are under development. These include a subunit vaccine formulated in MF59,[49] and a proteosome vaccine, containing outer membrane proteins of *Neisseria meningitidis*.[50] The proteosome vaccine has recently been evaluated in two active-comparator-controlled trials in healthy adults.[51] The study was conducted using a prototype monovalent A/Beijing/265/95 vaccine (H1N1). The vaccine was well tolerated and induced adequate levels of serum HI antibodies as well as a local sIgA response in the nasal cavity, the latter requiring a two-dose vaccination regimen.

Another intranasal influenza vaccine, NasalFlu®, containing native, non-detoxified, *E. coli* heat-labile enterotoxin (LT) as an adjuvant, was licensed in Switzerland in 2001. However, soon after it was introduced, the vaccine had to be withdrawn, because of an increased number of cases of Bell's palsy in association with the vaccination.[52] Although not formally established, it is likely that the occurrence of this complication was due to the toxic activity of the unmodified *E. coli* LT, the intranasal administration of the vaccine possibly allowing migration of the toxin to the brain. This incident underscores the numerous pitfalls that may be encountered in the development of a novel vaccine formulation, even after initial registration. In view of the NasalFlu® incident, the use of detoxified variants of CT and LT would appear to be preferable over the use of holotoxins. Non-toxic recombinant LTB retained potent adjuvant activity when administered intranasally to mice together with influenza subunit vaccine, inducing high levels of systemic IgG and local sIgA. The mice were completely protected from infection upon viral challenge.[20]

Novel approaches to vaccine virus production

While a number of experimental influenza vaccines are based on the use of antigen produced by recombinant DNA technology or even rely on the use of DNA itself as the immunogen, it is to be anticipated that in the years to come the primary basis for influenza vaccine production will remain the initial production of the influenza virus. Novel developments in this respect relate to the possibility of generating seed viruses by modern reverse genetics technology, and the use of mammalian cell-culture substrates for virus growth.

Reverse genetics

The technique of reverse genetics provides a powerful approach for the generation of recombinant influenza viruses with a desired combination of genes.[2] Previously, exchange of gene segments between different influenza viruses relied on deliberate reassortment, through co-infection of cells with two different viruses (see Chapter 3). In this manner, high-growth reassortants have been generated for decades, to provide seed viruses for production of vaccine virus in embryonated chicken eggs.

Influenza virus has long been refractory to reverse genetics approaches. The reason is that the viral genome consists of negative-sense RNA segments with unmodified 5' and 3' ends, which are replicated by a complex polymerase (see Chapter 2). However, robust systems for reverse genetics of influenza A and B viruses are now in place, which allow production of recombinant viruses from cloned cDNA. Initially, eight separate plasmids, encoding the respective viral RNA segments under control of the RNA polymerase I promoter, were used in combination with four plasmids for expression of the three polymerase subunits and the

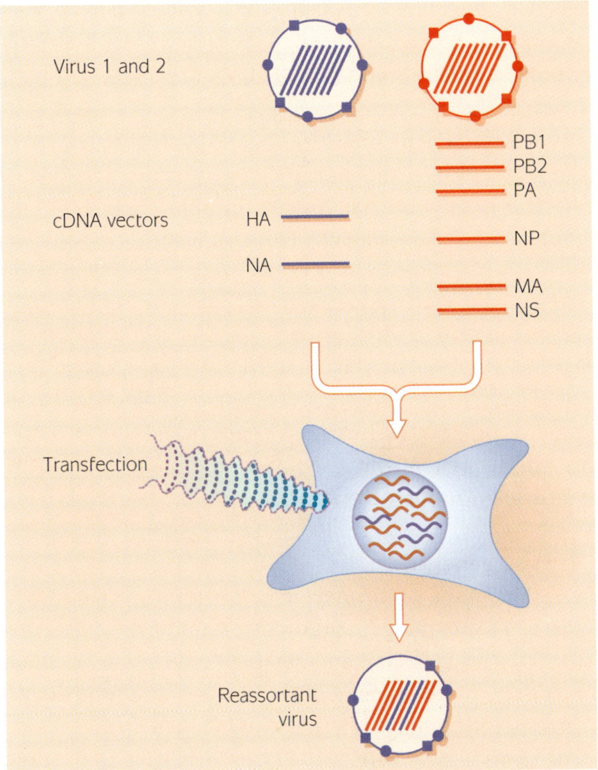

Figure 32. Schematic representation of the reverse-genetics technique for production of recombinant influenza viruses. A set of eight bidirectional transcription plasmids encoding the desired combination of influenza virus gene segments is transfected into cells, for example by electroporation. These plasmids produce messenger and as well as viral RNAs, ultimately resulting in the formation of new virus particles. Figure courtesy of Ron Fouchier, ErasmusMC, Rotterdam, the Netherlands.

nucleoprotein (NP), under control of the RNA polymerase II promoter, which are required for viral replication and transcription.[53] More recently, an eight-plasmid system has been developed in which the eight viral gene segments are flanked by an RNA polymerase I promoter on one side and an RNA polymerase II promoter on the other, such that negative-sense viral RNA and positive-sense mRNA can be produced from the same template (Figure 32).[54,55] Quite recently, a novel improved reverse genetics system has been developed, in which the eight polymerase I transcription cassettes for production of the viral RNAs are combined on a single plasmid, while similarly the polymerase II cassettes for production of the three polymerase proteins are combined on one plasmid.[56] This approach permits a very

Figure 33. Modern facility for production of influenza vaccine using cell-culture technology. Courtesy of Solvay Pharmaceuticals, Weesp, the Netherlands.

efficient production of recombinant influenza A viruses with the desired set of gene segments in Vero cells.

Reverse genetics techniques provide the opportunity to generate seed viruses for vaccine virus production either in eggs or in cell culture substrates (see below), regardless of the pathogenicity of the original wild-type virus strain. Indeed, the seed viruses for the first pandemic H5N1 vaccine candidates have been produced with this technology.

Cell-culture vaccines

Influenza virus can be propagated efficiently on a number of cell-culture substrates. Currently, three mammalian cell lines are being developed for the production of influenza vaccine virus.[57,58] Two of these, MDCK and Vero cells, are dog and monkey kidney cells, respectively. The third, the Per.C6 cell line, originates from a human embryonic retina cell.[59] Among these cell lines, Vero cells in particular have a long history in human vaccine production. For example, poliovirus vaccine is produced on Vero cells. Two inactivated cell-culture influenza vaccines, one produced on MDCK cells, the other on Vero cells, have already been licenced in a European Reference Member State (RMS) within the context of the Mutual Recognition Procedure (MRP) for the licensing of vaccines within Europe. Figure 33 shows a state-of-the-art facility for cell-culture production of influenza vaccine.

Extensive clinical evaluation of an inactivated influenza subunit vaccine, produced from virus grown on MDCK cells, has shown that it is safe and as immunogenic as subunit vaccine produced in eggs. Also, MDCK- and egg-derived vaccines show comparable reactogenicity.[60,61] Likewise, a Vero-produced whole virus vaccine has been found to be safe and immunogenic in clinical trials, although more people with a feverish reaction were observed. Because of this observation, the company involved decided in December 2004 to put further clinical trials planned for 2005 on hold, awaiting analysis of the data.

Clearly, cell-culture-based vaccine virus production provides a great deal of flexibility. Vaccine production is less vulnerable because it does not depend on the availability of embryonated chicken eggs. This is of particular importance in the case of a pandemic involving a bird-flu virus, which would hamper the egg supply for vaccine production, as was already the case in the H7N7 bird-flu outbreak in the Netherlands in 2003. Cell-culture-based virus production may be scaled up at short notice, such that large vaccine demands can be met as much as possible. Accordingly, the WHO has recognized the need for development of cell-culture techniques for the production of influenza vaccines, supplemental to or as an alternative for egg-based vaccine production.[62] In addition, efficient propagation of virus on cultured cells does not require the generation of classical high-growth reassortants. This is another potential time-saving advantage of cell-culture-based vaccine production. Furthermore, virus produced on mammalian cells retains the antigenic characteristics of the original virus isolates, whereas during virus propagation in eggs, variants with an altered antigenicity may be selected for. Finally, growth of vaccine virus in cell culture excludes the presence of contaminants that may be present in egg-grown vaccines, notably ovalbumin or other egg proteins causing occasional allergic reactions, or bacterial endotoxin derived from infected eggs.

Other ways of producing vaccine antigen

Influenza subunit vaccine can also be prepared without the prior growth of the entire influenza virus. For example, HA and/or NA may be produced in insect cells using a baculovirus expression vector. An experimental recombinant HA subunit vaccine produced in this manner has been evaluated in Phase I and II clinical trials with promising results.[63] In a placebo-controlled field efficacy study in 460 healthy subjects, the vaccine was found to be over 85% efficacious against culture-positive influenza illness. Although the investigators speculate that these results may be better than those obtained with current licensed inactivated vaccines, no such vaccine was included in the trial to substantiate this prediction. As discussed for the LAIVs above, a potential advantage of a new vaccine compared to licensed vaccines, needs scientific demonstration and confirmation. Only direct comparative trials between new vaccines and licensed vaccines will provide credible scientific evidence for superiority.[44]

Pandemic vaccines

In case of a pandemic, the whole world population will be vulnerable to infection. The WHO has estimated that 20–35% of the population may be affected.[64] Vaccines will be the most important medical intervention to contain the pandemic. However, it will take at least 4–6 months after the WHO has announced the onset of the pandemic (see Chapter 3) before the first pandemic vaccine doses will become available for distribution. There certainly will be vaccine shortages at the onset of the pandemic and most likely for a substantial period of time during its course.[65] At the early stage, the control and containment of the pandemic should come from general measures, and the use of antiviral agents as described in Chapter 7.

The WHO has recommended that countries finalize advance purchase agreements for pandemic vaccines with one or more vaccine manufacturers.[66] Countries with such agreements will be the first beneficiaries of the scarce volumes of pandemic vaccine. Only a few countries have such agreements in place at the moment. Although there is a general moral feeling that there should be an "equitable" distribution of scarce pandemic vaccines, there is currently no mechanism in place in the world on how to achieve this. In the absence of sufficient pre-pandemic purchase contracts, the distribution of scarce pandemic vaccines will at best follow the principle of "pro rata" distribution. Thus, countries without an advance vaccine purchase agreement would receive the number of pandemic vaccine doses according to the level of their seasonal influenza vaccine usage. Even though it is questionable whether, under conditions of an international emergency, a "pro rata" distribution will be followed, this principle was used during a pandemic exercise, where the degree of preparedness of a number of EU countries was tested, as discussed in Chapter 3.

In November 2005, the WHO, FAO, OIE and World Bank organized a unique global meeting on pandemic influenza in Geneva with all international stakeholders, where key issues were discussed.[67] Earlier recommendations by the WHO to improve seasonal influenza prevention by increase of annual vaccine uptake[68] were reinforced. As for any commercial product in free-economy countries, the production capacity for influenza vaccines will be matched over time with demand. Indeed, the current global production capacity for seasonal trivalent influenza vaccines matches the current level of demand. An increased annual usage of seasonal influenza vaccine would be a natural incentive for vaccine manufacturers to make the necessary investments to enlarge their production facilities. Increased global production capacity will obviously also enhance the volume of pandemic vaccine which can be produced when needed and therefore be an effective measure to improve the level of national and international pandemic preparedness. It would also contribute to the principle of "equitable distribution".

Pandemic vaccine candidates

There are several specific issues which make pandemic vaccines different from epidemic vaccines. These include: (i) the production of seed virus based on a highly pathogenic virus subtype; (ii) the need to include an adjuvant in the vaccine formulation so as to minimize the dose of antigen required for induction of a protective immune response and thus to maximize an equitable distribution of the inevitably scarse volumes of vaccine; and (iii) the implementation of flexible vaccine production methodologies allowing a very swift response to an emerging pandemic, such that as much vaccine as possible can be produced in the shortest possible time.

Seed virus production

Because of the high pathogenicity of a potential pandemic virus, specific safety requirements have been formulated for laboratories and factories to work with such viruses.[69] To allow the production of vaccine against the highly pathogenic avian H5N1 virus strain, the WHO has made a non-pathogenic H5N1 seed virus, using the reverse genetics technology presented above.

Adjuvants are required in pandemic vaccines

A recent study with a non-adjuvanted candidate pandemic vaccine has demonstrated that a total of approximately 12 times a regular antigenic dose would be needed to stimulate an adequate seroresponse.[70] This study confirmed the need for adjuvant formulations for potential pandemic vaccines. Indeed, a follow-up study, using the same vaccine but adjuvanted with aluminium reduced the dose level needed for an adequate seroresponse considerably.[71]

Because of the worldwide and long-term experience with aluminium-adjuvanted vaccines for other diseases, current pandemic vaccine research concentrates on this adjuvant.[72] Because, as indicated above, MF59 is used in the formulation of a commercially available adjuvanted influenza vaccine, this adjuvant is also being explored for potential application in pandemic vaccines. Other adjuvants may also be of interest, but due to the lack of experience with many of them, the clinical development to prove safety and efficacy of (pandemic) vaccines containing these adjuvants will be much more time-consuming and costly.

An MF59-adjuvanted candidate pandemic vaccine formulation has been demonstrated to have a clear-cut advantage over non-adjuvanted vaccine. In recent studies among volunteers, involving an experimental MF59-adjuvanted subunit vaccine against the H5N3[13] and H5N1[14] viruses, protective antibody titres could be achieved at a much lower dose in the volunteers receiving adjuvanted vaccine compared to the non-adjuvanted vaccine. Similar results were found for non-adjuvanted split and whole virus vaccine formulations with H9N2[15,16] and H5N1[70,71] antigens. Using a two-dose regimen, with the aluminium-adjuvanted vaccine formulations only, acceptable antibody titres could be demonstrated at normal or even lower vaccine doses. During a WHO meeting in Geneva in November 2005, all companies provided their experience with pandemic vaccine formulations and doses needed for a protective seroresponse.[72]

In conclusion, recent research with potential pandemic vaccine candidates has clearly shown both that unrealistically high doses of non-adjuvanted vaccine formulations are needed to induce an adequate level of protective antibodies, and that the use of adjuvants can reduce the required dose level considerably. Because of the demonstrated importance of adjuvants for the development of inactivated pandemic vaccines, research should be encouraged in this field.

Live attenuated pandemic vaccines

LAIVs, discussed above, are also being considered as potential pandemic vaccines. They may be interesting candidates because of their expected high level of immunogenicity for immunologically naive persons and their ability to rapidly induce immunity, i.e. within 10 days after vaccination.[4] However, there is also a major concern of reassortment of the live virus vaccine strain with a circulating human strain. This would have negative implications on the use of such a vaccine, particularly if the vaccine were used, while the pandemic would not materialize.[4] Although using every available option to develop vaccines that may be used for an influenza pandemic is critical,[4] the selection of the actual vaccine formulations to be used as pandemic vaccines should carefully balance the real and theoretical risk–benefit ratio at the individual as well as the societal level.

Registration issues related to pandemic vaccines

Once the optimal dose and dosing regimen to balance efficacy and production and supply potential for each of the candidate "mock-up" pandemic vaccines has been established, the candidate vaccines need to be safety tested and clinical studies need to be done for registration purposes. The European Medicines Agency (EMEA) has issued guidelines for the pre-registration of pandemic vaccines.[73] The pre-registration process requires the submission of a core dossier which contains the safety and immunogenicity data of the mock-up pandemic vaccine. The vaccine should be formulated and produced like the real pandemic vaccine, which may have to be produced years later. Once the WHO has issued the recommendation to start production of pandemic vaccine (Pandemic Phase 6;[64] see Table 4), and hence to stop the production of regular influenza vaccine, the real pandemic vaccine will be produced within the shortest possible time frame and distribution will be allowed before any clinical studies are done with this real pandemic vaccine. Additional clinical studies and a pharmacovigilance programme for each of the distributed pandemic vaccines will be required during the course of the pandemic.[73] In September 2005, the first core dossier for an aluminium-adjuvanted mock-up pandemic vaccine has been submitted to the EMEA for approval.

Most countries do not yet have a formal registration procedure in place for pandemic vaccines. In March 2006, the WHO is organizing an international regulatory meeting to co-ordinate a global approach to this important aspect of pandemic preparedness. Pandemic vaccines can only be effectively used if they can be distributed all over the world without delay due to differences in national registration procedures.

Is the world sufficiently prepared for the next pandemic?

At this point in time, the world is not yet sufficiently prepared to adequately manage a potential future influenza pandemic. Global viral and disease surveillance programmes for early detection, improved use of seasonal influenza vaccines, pandemic vaccine research, and development of improved influenza vaccines and antiviral agents, are key short- and long-term activities to increase the level of global pandemic preparedness. The world health community, coordinated by the WHO and with the help of the World Bank, is increasing its capacity for the necessary global and national surveillance programmes.

National health authorities in several countries are putting programmes in place to fully implement annual influenza prevention and control programmes according to existing recommendations, and possibly expanding these recommendations. Some have finalized national pre-pandemic purchase agreements for pandemic vaccines and antiviral stockpiling. On the other hand, many national health authorities still need to take such decisions for immediate action. Health-care professionals need to take care that at least their at-risk patients will be immunized each year against influenza.

Although it is clear that much still needs to be done to contain the disastrous societal effects of a next pandemic,[64] the current global efforts by WHO, health authorities, non-governmental organizations and industry alike give rise to optimism that our capabilities for effective management of a future pandemic are growing fast and that, when the next pandemic strikes, the world will be prepared as well and as reasonably as possible.

References

1. Nicholson KL, Wood JM, Zambon M. Influenza. *Lancet* 2003; **362**: 1733–1745.

2. Kemble G, Greenberg H. Novel generations of influenza vaccines. *Vaccine* 2003; **21**: 1789–1795.

3. Rimmelzwaan GF, Osterhaus ADME. Influenza vaccines: new developments. *Curr Opin Pharmacol* 2001; **1**: 491–496.

4. Luke CJ, Subburao K. Vaccines for pandemic influenza. *Emerg Infect Dis* 2006; **12**: 66–72.

5. Palese P. Improvement of influenza virus vaccines. *Emerg Infect Dis* 2006; **12**: 61–65.

6. Deng Y, Jing Y, Campbell AE *et al.* Age-related impaired type 1 T cell responses to influenza: reduced activation *ex vivo*, decreased expansion in CTL culture *in vitro*, and blunted response to influenza vaccination *in vivo* in the elderly. *J Immunol* 2004; **172**: 3437–3446.

7. Provinciali M, Di Stefano G, Muzzioli M *et al.* Impaired antibody response to influenza vaccine in institutionalized elderly. *Ann NY Acad Sci* 1994; **717**: 307–314.

8. Webster RG. Immunity to influenza in the elderly. *Vaccine* 2000; **18**: 1686–1689.

9. Boon AC, Fringuelli E, Graus YM *et al.* Influenza A virus specific T cell immunity in humans during aging. *Virology* 2002; **299**: 100–108.

10. Voordouw B. Influenza vaccination in community dwelling elderly persons. PhD thesis, Erasmus University, Rotterdam, 2005. ISBN 90-8559-103-1.

11. Thomas PG, Keating R, Hulse-Post DJ, Doherty PC. Cell-mediated protection in influenza infection. *Emerg Infect Dis* 2006; **12**: 48–54.

12. Neuzil KM, Edwards KM. Influenza vaccines in children. *Semin Pediatr Infect Dis* 2002; **13**: 174–181.

13. Nicholson KG, Colegate AE, Podda A *et al.* Safety and antigenicity of non-adjuvanted and MF59-adjuvanted influenza A/Duck/Singapore/97 (H5N3) vaccine: a randomised trial of two potential vaccines against H5N1 influenza. *Lancet* 2001; **357**: 1937–1943.

14. Stephenson I, Bugarini R, Nicholson KG *et al.* Cross-reactivity to highly pathogenic avian influenza H5N1 viruses after vaccination with nonadjuvanted and MF59-adjuvanted influenza A/Duck/Singapore/97 (H5N3) vaccine: a potential priming strategy. *J Infect Dis* 2005; **191**: 1210–1215.

15. Hehme N, Engelmann H, Kuenzel W, Neumeier E, Saenger R. Immunogenicity of a monovalent, aluminum-adjuvanted influenza whole virus vaccine for pandemic use. *Virus Res* 2004; **103**: 163–171.

16. Stephenson I, Nicholson KG, Gluck R *et al.* Safety and antigenicity of whole virus and subunit influenza A/Hong Kong/1073/99 (H9N2) vaccine in healthy adults: phase I randomised trial. *Lancet* 2003; **362**: 1959–1966.

17. Stephenson I, Nicholson KG, Wood JM, Zambon MC, Katz JM. Confronting the avian influenza threat: vaccine development for a potential pandemic. *Lancet Infect Dis* 2004; **4**: 499–509.

18. Podda A. The adjuvanted influenza vaccines with novel adjuvants: experience with the MF59-adjuvanted vaccine. *Vaccine* 2001; **19**: 2673–2680.

19. Snider DP. The mucosal adjuvants activities of ADP-ribosylating bacterial enterotoxins. *Crit Rev Immunol* 1995; **15**: 317–348.

20. De Haan L, Verweij WR, Holtrop M *et al.* Nasal or intramuscular immunization of mice with influenza subunit antigen and the B subunit of *Escherichia coli* heat-labile toxin induces IgA- or IgG-mediated protective mucosal immunity. *Vaccine* 2001; **19**: 2898–2907.

21. Babai I, Barenholz Y, Zakay-Rones Z *et al.* A novel liposomal influenza vaccine (INFLUSOME-VAC) containing hemagglutinin –neuraminidase and IL-2 or GM-CSF induces protective anti-neuraminidase antibodies cross-reacting with a wide spectrum of influenza A viral strains. *Vaccine* 2001; **20**: 505–515.

22. Ben-Yehuda A, Joseph A, Barenholz Y *et al.* Immunogenicity and safety of a novel IL-2-supplemented liposomal influenza vaccine (INFLUSOME-VAC) in nursing-home residents. *Vaccine* 2003; **21**: 3169–3178.

23. Stegmann T, Morselt HW, Booy FP, Van Breemen JF, Scherphof G, Wilschut J. Functional reconstitution of influenza virus envelopes. *EMBO J* 1987; **6**: 2651–2659.

24. Bungener L, Serre K, Bijl L *et al.* Virosome-mediated delivery of protein antigens to dendritic cells. *Vaccine* 2002; **20**: 2287–2295.

25. Bungener L, Huckriede A, de Mare A *et al.* Virosome-mediated delivery of protein antigens *in vivo*: efficient induction of class I MHC-restricted cytotoxic T lymphocyte activity. *Vaccine* 2005; **23**: 1232–1241.

26. Huckriede A, Bungener L, Stegmann T *et al.* The virosome concept for influenza vaccines. *Vaccine* 2005; **23** (Suppl 1): S26–S38.

27. Glück R, Moser C, Metcalfe IC. Influenza virosomes as an efficient system for adjuvanted vaccine delivery. *Expert Opin Biol Ther* 2004; **4**: 1139–1145.

28. De Bruijn IA, Nauta J, Gerez L *et al.* Virosomal influenza vaccine: a safe and effective influenza vaccine with high efficacy in elderly and subjects with low pre-vaccination antibody titers. *Virus Res* 2004; **103**: 139–145.

29. Kanra G, Marchisio P, Feiterna-Sperling C *et al.* Comparison of immunogenicity and tolerability of a virosome-adjuvanted and a split influenza vaccine in children. *Pediatr Infect Dis J* 2004; **23**: 300–306.

30. Ruf BR, Colberg K, Frick M *et al.* Open, randomized study to compare the immunogenicity and reactogenicity of an influenza split vaccine with an MF59-adjuvanted subunit vaccine and a virosome-based subunit vaccine in elderly. *Infection* 2004; **32**: 191–198.

31. De Bruijn IA, Nauta L, Gerez L *et al.* The virosomal vaccine Invivac: immunogenicity and tolerability compared to an adjuvanted influenza vaccine (Fluad) in elderly subjects. *Vaccine* 2006; in press.

32. Rimmelzwaan GF, Baars M, Van Amerongen G *et al.* A single dose of an ISCOM influenza vaccine induces long-lasting protective immunity against homologous challenge infection but fails to protect Cynomolgus macaques against distant drift variants of influenza A (H3N2) viruses. *Vaccine* 2001; **20**: 158–163.

33. Voeten JT, Rimmelzwaan GF, Nieuwkoop NJ *et al.* Introduction of the hemagglutinin transmembrane region in the influenza virus matrix protein facilitates its incorporation into ISCOM and activation of specific CD8(+) cytotoxic T lymphocytes. *Vaccine* 2000; **19**: 514–522.

34. Rimmelzwaan GF, Nieuwkoop N, Brandenburg A *et al.* A randomized, double blind study in young healthy adults comparing cell-mediated and humoral immune responses

induced by influenza ISCOM vaccines and conventional vaccines. *Vaccine* 2000; **19**: 1180–1187.

35. Webster RG. DNA vaccination: a potential future strategy. In: Nicholson KG, Webster RG, Hay AJ, editors. *Textbook of Influenza*. Blackwell Science, 1998; pp. 410–421.

36. Ulmer JB. Influenza DNA vaccines. *Vaccine* 2002; **20** (Suppl 2): S74–S76.

37. Neirynck S, Deroo T, Saelens X *et al*. A universal influenza A vaccine based on the extracellular domain of the M2 protein. *Nat Med* 1999; **5**: 1157–1163.

38. Heinen PP, Rijsewijk FA, de Boer-Luijtze EA *et al*. Vaccination of pigs with a DNA construct expressing an influenza virus M2-nucleoprotein fusion protein exacerbates disease after challenge with influenza A virus. *J Gen Virol* 2002; **83**: 1851–1859.

39. Waldman RH, Bond JO, Levitt LP *et al*. An evaluation of influenza immunization: influence of route of administration and vaccine strain. *Bull World Health Org* 1969; **41**: 543–548.

40. Ghendon Y. Cold-adapted, live influenza vaccines developed in Russia. In: Nicholson KG, Webster RG, Hay AJ, editors. *Textbook of Influenza*. Blackwell Science, 1998; pp. 391–399.

41. Glezen WP. Cold-adapted, live attenuated influenza vaccine. *Expert Rev Vaccines* 2004; **3**: 131–139.

42. Mendelman PM, Cordova J, Cho I. Safety, efficacy and effectiveness of the influenza virus vaccine, trivalent, types A and B, live, cold-adapted (CAIV-T) in healthy children and healthy adults. *Vaccine* 2001; **19**: 2221–2226.

43. Belshe RB. Current status of live attenuated influenza virus vaccine in the US. *Virus Res* 2004; **103:** 177–185.

44. Beyer WEP, Palache AM, De Jong JC, Osterhaus ADME. Cold-adapted live influenza vaccine versus inactivated vaccine: systemic vaccine reactions, local and systemic antibody response, and vaccine efficacy. A meta-analysis. *Vaccine* 2002; **20**: 1340–1353.

45. Centers for Disease Control and Prevention. Prevention and control of influenza. *Recommendations of the Advisory Committee on Immunization Practices (ACIP)* 2004; MMWR 28(RR-6).

46. Palker T, Kisele I, Johnston K *et al.* Protective efficacy of intranasal cold-adapted influenza A/New Caledonia/20/99 (H1N1) vaccines comprised of egg- or cell culture-derived reassortants. *Virus Research* 2004; **105**: 183–194.

47. Watanabe T, Watanabe S, Neumann G, Kida H, Kawaoka Y. Immunogenicity and protective efficacy of replication-incompetent influenza virus-like particles. *J Virol* 2002; **76**: 767–773.

48. Watanabe T, Watanabe S, Kida H, Kawaoka Y. Influenza A virus with defective M2 ion channel activity as a live vaccine. *Virology* 2002; **299**: 266–270.

49. Boyce TG, Hsu HH, Sannella EC *et al.* Safety and immunogenicity of adjuvanted and unadjuvanted subunit influenza vaccines administered intranasally to healthy adults. *Vaccine* 2000; **19**: 217–226.

50. Plante M, Jones T, Allard F *et al.* Nasal immunization with subunit proteosome influenza vaccines induces serum HAI, mucosal IgA and protection against influenza challenge. *Vaccine* 2001; **20**: 218–225.

51. Treanor J, Nolan C, O'Brien D *et al.* Intranasal administration of a proteosome-influenza vaccine is well-tolerated and induces serum and nasal secretion influenza antibodies in healthy human subjects. *Vaccine* 2006; **24**: 254–262.

52. Mutsch M, Zhou W, Rhodes P *et al.* Use of the inactivated intranasal influenza vaccine and the risk of Bell's palsy in Switzerland. *New Engl J Med* 2004; **350**: 896–903.

53. Neumann G, Watanabe T, Ito H *et al.* Generation of influenza A viruses entirely from cloned cDNAs. *Proc Natl Acad Sci USA* 1999; **96**: 9345–9350.

54. Hoffmann E, Neumann G, Kawaoka Y *et al.* A DNA transfection system for generation of influenza A virus from eight plasmids. *Proc Natl Acad Sci USA* 2000; **97**: 61086113.

55. De Wit E, Spronken MI, Bestebroer TM *et al.* Efficient generation and growth of influenza virus A/PR/8/34 from eight cDNA fragments. *Virus Res* 2004; **103**: 155–161.

56. Neumann G, Fujii K, Kino Y *et al.* An improved reverse genetics system for influenza A virus generation and its

implications for vaccine production. *Proc Natl Acad Sci USA* 2005; **102**: 16825–16829.

57. Brands R, Visser J, Medema J *et al.* InfluvacTC: a safe Madin Darby Canine Kidney (MDCK) cell culture-based influenza vaccine. *Dev Biol Stand* 1999; **98**: 93–100.

58. Kistner O, Barrett PN, Mundt W, Reiter M, Schober-Bendixen S, Dorner F. Development of a mammalian cell (Vero) derived candidate influenza virus vaccine. *Vaccine* 1998; **16**: 960–968.

59. Pau MG, Ophorst C, Koldijk MH *et al.* The human cell line PER.C6 provides a new manufacturing system for the production of influenza vaccines. *Vaccine* 2001; **19**: 2716–2721.

60. Palache AM, Scheepers HS, De Regt V *et al.* Safety, reactogenicity and immunogenicity of Madin Darby Canine Kidney cell-derived inactivated influenza subunit vaccine. A meta-analysis of clinical studies. *Dev Biol Stand* 1999; **98**: 115–125.

61. Halperin SA, Smith B, Mabrouk T *et al.* Safety and immunogenicity of a trivalent, inactivated, mammalian cell culture-derived influenza vaccine in healthy adults, seniors, and children. *Vaccine* 2002; **20**: 1240–1247.

62. World Health Organization. Cell culture as a substrate for the production of influenza vaccines: Memorandum from WHO meeting. *Bull World Health Org* 1995; **73**: 431–435.

63. Lakey DL, Treanor JJ, Betts RF *et al.* Recombinant baculovirus influenza A hemagglutinin vaccines are well tolerated and immunogenic in healthy adults. *J Infect Dis* 1996; **174**: 838–841.

64. World Health Organization. WHO Global Influenza Preparedness Plan. The role of WHO and recommendations for national measures before and during pandemics. Geneva, March 2005. Document WHO/CDS/CSR/GIP/2005.5. (www.who.int/csr/resources/publications/influenza/GIP_2005_5Eweb.pdf)

65. Stöhr K, Esveld M. Will vaccines be available for the next influenza pandemic? *Science* 2004; **306**: 2195–2196.

66. World Health Organization. Guidelines on the use of vaccines and antivirals during influenza pandemics. Geneva, 2004. Document WHO/CDS/CSR/RMD/2004.8. (www.who.int/csr/resources/publications/influenza/en/11_29_01_A.pdf)

67. World Health Organization. Meeting on avian influenza and human pandemic influenza. Geneva, 7–9 November 2005. (www.who.int/mediacentre/events/2005/avian_influenza/en/index.html)

68. World Health Organization. Responding to the avian influenza pandemic threat. Recommended strategic actions. Geneva, 2005. Document WHO/CDS/CSR/RMD/2005.8. (www.who.int/csr/resources/publications/influenza/WHO_CDS_CSR_GIP_2005_8/en/index.html)

69. World Health Organization. WHO biosafety risk assessment and guidelines for the production and quality control of human influenza pandemic vaccines. Annex 5. Adopted by the 56th meeting of the WHO Expert Committee on Biological Standardization, 24–28 October 2005. (www.who.int/biologicals/publications/ECBS%202005%20Annex%205%20Influenza.pdf)

70. Lambert LC. *Update on NIH H5N1 Vaccine Trials: DMID 04-063.* National Vaccine Advisory Committee November 29, 2005.

71. Perez M. Grippe aviaire: un vaccin encourageant. *Le Figaro* December 15th, 2005.

72. World Health Organization. WHO meeting on development and evaluation of influenza pandemic vaccines. Geneva, 2–3 November 2005. (www.who.int/vaccine_research/diseases/influenza/mtg_021205/en/index.html; see also www.ifpma.com)

73. European Medicines Agency (EMEA). Guideline on submission of marketing authorisation application for pandemic influenza vaccines through the centralised procedure. April 2004. Document EMEA/CPMP/VEG/4986/03. (www.emea.eu.int/pdfs/human/vwp/498603en.pdf) (www.emea.eu.int/pdfs/human/vwp/471703en.pdf)

Appendix 1 – Useful Websites

World Health Organization
www.who.int/csr/disease/influenza

US Centers for Disease Control
www.cdc.gov/flu/

US National Institute of Allergy and Infectious Diseases
www.niaid.nih.gov/publications/flu.htm

Center for Infectious Disease Research and Policy, University of Minnesota, USA
www.cidrap.umn.edu

European Scientific Working Group on Influenza
www.eswi.org

European Influenza Surveillance Scheme
www.eiss.org

European Agency for the Evaluation of Medicinal Products
www.emea.eu.int/

European Centre for Disease Prevention and Control
www.europa.eu.int/comm/health/ph_overview/strategy/ecdc/ecdc_en.htm

Asia-Pacific Advisory Committee on Influenza
www.apaci-flu.com/apaci/

International Federation of Pharmaceutical Manufacturers and Associations
www.ifpma.org

European Vaccine Manufacturers
www.evm-vaccines.org

Appendix 2 – WHO Global Agenda for Influenza Surveillance and Control[1]

Objectives of the Global Agenda

- Provide impartial and prioritized guidance to all parties on research and development and national/global action for influenza control
- Support co-ordination of action for influenza control and surveillance
- Support implementation of identified priorities
- Support advocacy and fund raising

The Global Agenda has been developed for all those involved in activities to:

- Reduce morbidity and mortality from annual influenza epidemics
- Prepare for the next influenza pandemic
- Co-ordinate national and international action in influenza surveillance and control
- Advocate and raise funds

Major Themes in the Global Agenda

1. Improvement in the quality and coverage of virological and epidemiological influenza surveillance
2. Improvement in the understanding of health and economic burden of influenza, including benefits from epidemic control and pandemic preparedness
3. Expansion in the use of existing vaccines, particularly in developing countries and in high-risk groups and acceleration in the introduction of new vaccines
4. Increase in national and global epidemic and pandemic preparedness, including vaccine and pharmaceutical supplies

Priority activities and key actions to reduce morbidity and mortality from annual influenza epidemics and to prepare for the next pandemic

A. Strengthen virological and epidemiological surveillance nationally and internationally

i. Enhance and integrate virological and disease surveillance
ii. Expand virological and disease surveillance
iii. Expand animal influenza surveillance and integrate with human influenza surveillance
iv. Improve data management, utilization and exchange

B. Increase knowledge on the health and economic burden of influenza

i. Strengthen the capacity in epidemiological and statistical techniques for studies on influenza disease burden
ii. Evaluate the clinical and economic burden of disease in countries where there is no recognition of influenza or no control policies are in place
iii. Re-evaluate the clinical and economic burden of influenza in countries where influenza control policies are in place

C. Increase influenza vaccine usage

i. Encourage assessment of disease burden and cost-effectiveness analyses
ii. Encourage countries to establish national policies and set immunization targets
iii. Promote awareness among policy makers, health care providers, and the public
iv. Encourage countries to identify and develop effective strategies for vaccine delivery
v. Develop and implement methods for the measurement and feedback of the progress of national and local programmes

D. Accelerate national and international action on pandemic preparedness

i. Increase awareness of the need for pandemic planning
ii. Accelerate the development and implementation of national pandemic plans
iii. Enhance the utilization of influenza vaccine and antivirals in the inter-pandemic period
iv. Develop strategies for the utilization of vaccines and antivirals and secure adequate supplies for a pandemic
v. Advocate research on pandemic viruses, vaccines, antivirals and other control measures

The Global Agenda on Influenza Surveillance and Control was developed and agreed by consensus by the participants of the WHO Consultation on Global Priorities in Influenza, World Health Organization, Geneva, Switzerland, 6–7 May 2002.

[1] The full text of the WHO Global Agenda for Influenza Surveillance and Control can be found at:

- www.who.int/csr/disease/influenza/globalagenda/en/
- www.who.int/csr/disease/influenza/csrinfluenzaglobalagenda/en/

Appendix 3 – Position Paper on Influenza Vaccines

Extracts from a World Health Organization document.[1]

Data on vaccination coverage show that even in industrialized countries, large proportions of the population at risk do not receive the influenza vaccine. WHO therefore encourages initiatives to raise awareness of influenza and influenza vaccination among health care workers and the public, and encourages setting of national targets for vaccination coverage.

The justification for vaccine use

During influenza epidemics, attack rates of 1–5% are most commonly observed, but the attack rate may reach 40–50% or more among elderly persons in institutions and in other high-risk groups. At least in western communities, bacterial complications such as pneumonia are frequently associated with influenza; the total annual excess mortality during influenza epidemics is estimated at 7.5–23 per 100,000. Influenza poses a considerable economic burden both on society and the individual in terms of consumption of health care resources and lost productivity.

Internationally licensed influenza vaccines have proven to be efficacious and safe. During influenza outbreaks, appropriate vaccination may significantly reduce respiratory illness and sick-leave among healthy adults. More importantly, vaccination may reduce severe disease and premature death in the elderly and in persons with underlying ailments or disease (for details on vaccine efficacy, see below).

Antiviral drugs such as the M2 inhibitors (acting against type A virus) and the more recently developed neuraminidase inhibitors (acting against both type A and type B viruses) have been shown to be effective for treatment (and for some agents, prophylaxis) and are now available in many industrialized countries. Resistant mutants to both classes of antiviral agents have been detected, and antimicrobial resistance surveillance is important to assess the magnitude of this problem. Also, costs, occasional side effects and the likely limited availability of such drugs during major outbreaks highlight the role of vaccination as the primary preventive measure against influenza.

Influenza virus vaccines

The three types of inactivated influenza vaccine show comparable efficacy, but differ in terms of reactogenicity. Thus in 15–20% of vaccines, the whole virus vaccines cause local reactions lasting for 1–2 days. Such reactions appear to be more common in young children than in adults. Transient systemic reactions such as fever, malaise and myalgias may occur in a minority of vaccine recipients within 6–12 hours of the vaccination. Split vaccines and subunit vaccines show reduced systemic reactogenicity both in children and in adults as compared to whole virus preparations. Consequently, subunit and split-virus vaccines are more attractive, particularly for use in children.

WHO position on influenza vaccines

The main purpose of influenza vaccination is to avoid severe influenza and its complications. This paper is concerned mainly with epidemic influenza and the public health impact of yearly influenza vaccination. Authoritative information on pandemics can be found in the WHO influenza pandemic plan. Recommendations for the use of inactivated influenza vaccines and other preventive measures are published in the weekly epidemiological record.

Most of the widely licensed influenza vaccines are manufactured according to the quality requirements defined by WHO and have proven efficacious in the elderly and other groups at risk. If influenza vaccination of children is required, for example as a consequence of predisposing conditions, the vaccine will not interfere with diphtheria–tetanus–pertussis (DTP) or other childhood vaccines, possibly due at the same time. To reduce adverse effects, only split vaccines or subunit vaccines should be given to children. Influenza vaccine should not be given to children aged under 6 months, and those aged 6–35 months should only receive half the adult vaccine dose.

Ideally, when major outbreaks are expected, all individuals should have the opportunity to be vaccinated against influenza. However, limited health budgets and, at least initially, shortage of the appropriate vaccine may force health authorities to restrict influenza vaccine to groups at particular risk. The following priority is recommended:

1. Residents of long-term care facilities for the elderly and the disabled – they are considered at high risk of influenza and its complications
2. Elderly non-institutionalized individuals suffering from chronic conditions such as pulmonary or cardiovascular illness, metabolic diseases including diabetes mellitus and renal dysfunction, and various types of immunosuppression including persons with AIDS and transplant recipients
3. All adults and children aged over 6 months suffering from any of the conditions mentioned above
4. Individuals who are above a nationally defined age limit irrespective of other risk factors. Although the appropriate age for general vaccination may be considerably lower in countries with poor living conditions, most countries define the limit age >65 years
5. Other groups defined on the basis of national data
6. Health care workers in regular, frequent contact with high-risk persons
7. Household contacts of high-risk persons.

Pregnant women who will be in their second or third trimester by the start of the influenza season and who are likely to be exposed are advised to consider vaccination in careful consultation with a competent healthcare provider. From a societal perspective, there are good arguments for influenza vaccination of children and healthy adults. Where adequate vaccine supplies are available, vaccination of the general public may be considered. However, the implementation of large-scale influenza vaccination programmes for these groups requires further evaluation of cost-effectiveness and cannot be generally recommended until firm data are presented. Nevertheless, persons who provide essential community services should be considered for vaccination. In some developed countries, companies find it economically justifiable to offer vaccination to their employees.

Although the WHO global influenza surveillance network has proven to be a reliable and successful system, it is important to increase worldwide coverage. Many countries are not included in the network, and in some large countries more than one centre is required. Surveillance is of particular importance in rural areas where potential animal hosts and humans live in close proximity, since it is in such areas that new viral recombinants are likely to originate.

[1] World Health Organization. Influenza vaccines. *Wkly Epidemiol Rec* 2002; **77**: 230–239.
(www.who.int/docstore/wer/77_27_52.html)

The relatively low uptake of influenza vaccines in most industrialized countries implies that significant proportions of the groups at risk of complications from influenza are not vaccinated. WHO strongly emphasizes the importance of raising the public consciousness of influenza and its complications as well as of the beneficial effects of influenza vaccination.

Appendix 4 – Avian Influenza: Assessing the Pandemic Threat

Extracts from a World Health Organization document.[1]

Foreword by Dr LEE Jong-wook, Director-General of the WHO

In the past, pandemics have announced themselves with a sudden explosion of cases which took the world by surprise. This time, we have been given a clear warning. During 2004, large parts of Asia experienced unprecedented outbreaks of highly pathogenic avian influenza, caused by the H5Na virus, in poultry. The virus crossed the species barrier to infect humans, with a high rate of mortality. Monitoring of the evolving situation, co-ordinated by WHO, has produced many signs that a pandemic may be imminent. This time, the world has an opportunity to defend itself against a virus with pandemic potential before it strikes.

Preparedness for a pandemic presents a dilemma: what priority should be given to an unpredictable but potentially catastrophic event, when many existing and urgent health needs remain unmet? In such a situation, it is useful to put together all known facts that can help us to see where we stand, what can happen, and what must be done.

The H5N1 virus has given us not only a clear warning but time to enhance preparedness. During 2004, concerns about the threat of a pandemic set in motion a number of activities, co-ordinated by WHO, that are leaving the world better prepared for the next pandemic, whenever it occurs and whichever virus causes it. Nonetheless, our highly mobile and interconnected world remains extremely vulnerable. No one can say whether the present situation will turn out to be another narrow escape or the prelude to the first pandemic of the 21st century. Should the latter event occur, we must not be caught unprepared.

The H5N1 outbreaks in 2004: a pandemic in waiting?

At some unknown time prior to 1997, the H5N1 strain of avian influenza virus began circulating in the poultry populations of parts of Asia, quietly establishing itself. Like other avian viruses of the H5 and H7 subtypes, H5N1 initially caused only mild disease with symptoms, such as ruffled feathers and reduced egg production, that escaped detection. After months of circulation in chickens, the virus mutated to a highly pathogenic form that could kill chickens within 48 hours, with a mortality approaching 100%. The virus first erupted in its highly pathogenic form in 1997, but did not appear again. Then, towards the end of 2003, H5N1 suddenly became highly and widely visible.

That finding grabbed the immediate attention of health experts. Of all viruses in the vast avian influenza pool, H5N1 is of particular concern for human health for two reasons. First, H5N1, though strictly an avian pathogen, has a documented ability to pass directly from birds to humans. Second, once in humans, H5N1 causes severe disease with very high mortality. These two features combine to make H5N1 of concern for a third and greater reason: its potential to ignite an especially severe pandemic.

A striking feature of the H5N1 Hong Kong outbreak was the presence of primary viral pneumonia in severe cases. When pneumonia occurs in influenza patients, it is usually a complication caused by a secondary bacterial infection. In the H5N1 cases, pneumonia was directly caused by the virus, did not respond to antibiotics, and was frequently rapidly fatal. With one exception, none of these patients had underlying disorders that could explain the severe course of the disease.

In January 2004, WHO officials were understandably on high alert for any signs that H5N1 might again cross the species barrier to cause disease in humans. On 5 January, Vietnamese health authorities informed the WHO office in Hanoi of an unusual cluster of severe respiratory disease in 11 previously healthy children hospitalized in

Hanoi. Of these patients, seven had died and two were in critical condition. Treatment with antibiotics produced no response, and a viral cause was suspected. Infection with the SARS virus was considered but did not seem likely. For unknown reasons, SARS tended to spare children, rarely causing severe illness, and was never considered a paediatric disease.

Fully aware of the risks, WHO activated its pandemic preparedness plan, alerted its network laboratories, and placed response teams on standby. WHO also mapped out a response plan with three objectives: to avert a pandemic, to control the outbreak in humans and prevent further cases, and to conduct the research needed to monitor the situation and improve preparedness, including the immediate development of a pandemic vaccine.

Lessons from past pandemics

Pandemics are always remarkable global events. Caused as they are by a highly contagious virus to which populations have little if any immunity, they benefit from almost universal susceptibility to infection. This gives them their distinctive features: they spread to all parts of the world very quickly, usually within less than a year, and cause illness in more than a quarter of the total population. It is this abrupt upsurge in illness, outstripping response capacity, that makes pandemics so disruptive, in addition to the excess mortality they invariable cause.

Actions in the face of pandemic threat

In the midst of all these unknowns, one epidemiological event is certain: health systems around the world will be confronted by a sudden and sharp increase in the demand for health care. The rapid global spread which has historically characterized pandemics will very likely be accelerated in today's highly mobile world. While the speed of information spread has no direct effect on morbidity and mortality, it may compromise response capacity if large populations within a country or geographical region are affected almost simultaneously. That situation would preclude the generous assistance so often provided during humanitarian crises in which only a single country or geographical region has been affected and the rest of the world is spared. Judging from past experiences with pandemics, good health systems and standards of care, high levels of sanitation and hygiene, and adequate resources may reduce mortality during a pandemic but cannot protect countries against the arrival and rapid spread of a highly contagious disease caused by a virus that will be largely or entirely foreign to the human immune system.

This mixture of unknowns and certainties creates a familiar but difficult public health dilemma: what priority should be given to preparedness for an inevitably recurring event of unpredictable timing and an outcome that is also unpredictable but could be catastrophic? Public health officials in a number of countries, faced with a chronic shortage of funds, must often regard preparedness for some future emergency as a luxury when viewed against the many other immediate and urgent infectious disease threats competing for resources.

The unpredictable nature of influenza viruses makes it impossible to know whether recent events will turn out to be another close call with a dangerous virus, or the prelude to the first pandemic of the 21 century. Should the latter event occur, the world will find itself warned far in advance, better prepared than at the start of 2004, yet still highly vulnerable.

[1] World Health Organization. Avian influenza: assessing the pandemic threat. 2005.
(www.who.int/csr/disease/influenza/WHO_CDS_2005_29/en)

Index

Notes

As influenza is the subject of this book, all index entries refer to influenza unless otherwise stated, *see* individual strains for more specific information. Page numbers followed by 'f' indicate figures: page numbers followed by 't' indicate tables or boxed material.